# Medicinal Plants and the Legacy of Richard E. Schultes

# Medicinal Plants and the Legacy of Richard E. Schultes

Proceedings of the Botany 2011 Symposium Honoring

Dr. Richard E. Schultes

Edited by
Bruce E. Ponman and Rainer W. Bussmann

The William L. Brown Center
at the Missouri Botanical Garden
P.O. Box 299
St. Louis, MO 63166-0299
USA

Impreso en GRAFICART SRL
San Martín 375 - Trujillo, Perú

ISBN-10: 0984841520
ISBN-13: 978-0-9848415-2-3

Hecho el Depósito Legal en la Biblioteca Nacional del Perú N° 2012-07285
Primera edición, Trujillo, Perú, mayo del 2012

Impreso en Perú
Printed in the Peru

Front cover photo by Richard E. Schultes.
Back cover image: Painting of R.E. Schultes in the Nash Laboratory, by Hannah Barrett. Courtesy of Harvard University Herbaria.

## Table of Contents

# Preface

Rainer W. Bussmann

Richard Evans Schultes has no doubt been one of the most important ethnobotanists of the 20th century. During decades of fieldwork, first in Mexico and then the Colombian Amazon, Schultes amassed a wealth of data later presented to the public in over 400 papers and dozens of books. Much of this work would, however, not have been possible without dedicated counterparts and students.

You hold now in your hands the proceedings of the first symposium to honor Richard Schultes, which took place in St. Louis, Missouri, on the 10th anniversary of his death. The idea for a symposium in Schultes' honor arose from a presentation on Richard Schultes' legacy that I was asked to give during the 2010 Botanical Society of America conference. While preparing the talk, I realized that even I, as a "second career ethnobotanist," knew very little about the man who formed so much of the discipline, that most students of ethnobotany knew even less, and that Schultes' collections were all but neglected in some attic at Harvard. The preservation of traditional knowledge for future generations is often seen as one of the main tasks of ethnobotanical fieldwork, and many of us are trying hard to achieve this goal when working with local communities, but in the case of Schultes' own work, the knowledge and legacy of our most iconic colleague was clearly getting lost.

Richard Schultes' former collaborators and students were clearly the best sources to give a perspective on their mentor's life and legacy. Thus, as many as possible were invited to the symposium to share their memories. Tracking down old collaborators proved difficult, however. From the beginning, it was clear that most contemporary colleagues of Richard Schultes had passed away. Hernando Garcia-Barriga, Schultes' great counterpart in Colombia, had died just a few years back. It was only with much dedication and a good portion of luck that Rodrigo Cámara-Leret, working on his Ph.D. in Colombia, was able to track down Isidoro Cabrera, who had been Richard Schultes' field assistant in the 1950s.

Although too frail in health to come to St. Louis, Isidoro Cabrera agreed to meet with

Figure 1. Richard Schultes, Hernando Garcia-Barriga, and Isidoro Cabrera at Río Apaporis, Colombia, 1951. Photo courtesy of Isidoro Cabrera.

Figure 2. Pedro Restrepo, Richard Schultes, and Isidoro Cabrera at the Hervario Nacional de Colombia, Bogota, 1952. Photo courtesy of Isidoro Cabrera.

Rodrigo at his home in Cali, Columbia, to share his memories about Richard Schultes, and to give a glimpse back to what ethnobotany fieldwork in 1950s Colombia looked like. Isidoro Canbrera accompanied Schultes and Garcia-Barriga to the Apaporis in 1951 (Fig. 1) and formed part of the team for years to come.

Most of the images in Isidoro Cabrera's home show little of the hardships the team encountered in the remote field and reflect much of Schultes' credo to appear in impeccable outfit whether he was in the field or in the laboratory (Fig. 2).

Schultes' main piece of equipment, a pith helmet, appears in almost every image (Figs. 3 and 4). But while the outfit of ethnobotanists in the 21st century has certainly changed, the field techniques, whether for collecting plants, or working with local informants, have stayed the same.

Richard Schultes, in fact, was the first ethnobotanist to step off the academic cloud and to respect and value the deep traditional knowledge his indigenous counterparts held.He also recognized that, without their help and collaboration, any ethnobotanical work would simply be impossible.

Although focusing on the investigators themselves, Isidoro Cabrera's pictures also convey the timeless beauty of the Amazon (Fig. 5) and its tremendous diversity, which despite 60 years of accelerated destruction by logging, mining, oil exploration, and (most recently) large scale agriculture and biofuel production, has endured to this day.

In this volume, Bob Bye, Mike Balick, Doel Soejarto, Jim Zarucchi, and Neil Schultes, some of Richard Schultes' most recognized

Figure 3. Isidoro Cabrera in the field, 1952. Photo courtesy of Isidoro Cabrera.

Figure 4. Isidoro Cabrera with Abel, son of Taiwano headman Abel, Río Tananari 1953. Photo courtesy of Isidoro Cabrera.

Figure 5. Isidoro Cabrera, Vaupes 1952. Photo courtesy of Isidoro Cabrera.

students, present their own perspective on their mentor, his life work, and his influence on their careers. Papers addressing Schultes' legacy and the current state of ethnobotany in the Amazon round out the picture.

What better way, however, to start a symposium to honor Richard Schultes, and start this volume, than the poetry of Isidoro Cabrera, written in Cali during the Spring of 2011 for the occasion:

Figure 6. Isidoro Cabrera, at his home in Cali, Colombia, 2011.
Photo courtesy of Rodrigo Cámara-Leret.

## A RICHARD EVANS SCHULTES, IN MEMORIAM EXCELSIOR

Incansable explorador
De la extensa Amazonia
Pues trabajó con amor
Estudiando su armonía.

Su vida expuso de veras
En los torrentosos ríos
De impetuosas "cachiveras"
En inviernos y en estíos.

Sabio de carácter fuerte
Sometido a muchas pruebas
Desafiando hasta la muerte
Dio a la ciencia plantas nuevas.
Codificó su experiencia
Con buenas observaciones
Para los hombres de ciencia
De nuevas generaciones.

Bajo la selva profunda
Que sus secretos encierra
Desconocida y fecunda
Cubre parte de la tierra.

El escribió su historia
Para siempre en el futuro
Y en su pedestal de gloria
Esta incólume y seguro.

# Medicinal Plants and the Legacy of Richard E. Schultes

# Reflections on Richard Evans Schultes, the Society for Economic Botany, and the Trajectory of Ethnobotanical Research

Michael J. Balick[1,2]

The life and work of Richard Evans Schultes—teacher, mentor and friend—resulted in a most extraordinary legacy that impacted an entire academic discipline as well as the tropical rainforest, where he spent so much of his time. His life and times have been brilliantly chronicled by Wade Davis in the biography *One River: Exploration and Discoveries in the Amazon Rain Forest* (1997). At this, the tenth anniversary of his passing in 2001, we celebrate, with this symposium, his life and academic contributions. Professor Schultes, as his students always called him with such great reverence, was a founder of The Society for Economic Botany in 1959 and edited our journal for 18 years, from 1962–1979. His multidisciplinary scientific research program began in 1936 with that well-known trip to Oklahoma and lasted far beyond his retirement from Harvard in 1985. Professor Schultes' mentorship of dozens of Ph.D. students during his career at Harvard, combined with his teaching, lecturing, and inspiration of tens of thousands of students throughout his long and distinguished career, did much to influence the trajectory of ethnobotany and economic botany. His academic philosophy and values were heavily influenced by his mentor, Oakes Ames. These values are deeply infused in the lives and professional activities of his numerous students, as can be seen from the presentations in this symposium.

A few weeks after returning to the United States from nearly a year of studies in the Costa Rican rainforest, I packed my bags once again and drove to Cambridge, Massachusetts. My purpose in traveling was to begin graduate studies at Harvard University, under the guidance of Professor Richard Evans Schultes. On my first visit to meet him, I climbed up the wrought iron staircase of the Botanical Museum where he had his laboratory, and marveled at the collection of glass flowers, carefully exhibited and curated under the watchful eye of my future mentor. This collection of exacting copies of plants was produced by Leopold Blaschka and his son, Rudolph, during five decades, and comprises over 830 species and 3,000 models. This priceless collection was commissioned in 1886 in order that professors could utilize the highest quality specimens, year around, in botany classes at the University.

Finally, at the top of the stairs, I knocked softly, and with a great degree of timidity, on the Professor's massive steel door. He was seated behind a pile of herbarium specimens collected from the Colombian Amazon, outfitted in grey flannel slacks, red suspenders, starched white shirt and a white laboratory coat. I was to learn that this was his uniform in Cambridge, as much as his pith helmet, khaki pants, and khaki shirt were his uniform in the Amazon. He greeted me with a fatherly smile and began to listen to my stories of life in the tropics, as he would do over the next two and a half decades. We exchanged notes on our travels, his of nearly half a century, mine embryonic by comparison. It was clear from the first meeting that I was in the presence of a most unusual person, a great thinker, and

---

1 A portion of this manuscript was published in the following obituary of Richard Evans Schultes: M.J. Balick. 2001. In gratitude Richard Evans Schultes 1915–2001. *Plant Talk* (July):34–35.

2 Director and Philecology Curator of Economic Botany, Institute of Economic Botany, The New York Botanical Garden, 2900 Southern Blvd., Bronx, NY 10458. Email: mbalick@nybg.org.

Figure 1. Young Richard Evans Schultes during his fieldwork investigating peyote in Oklahoma. Photo courtesy of R.E. Schultes.

an individual who viewed life as an opportunity to carry out the dreams of his early youth—to explore the most remote parts of the Amazon Valley, as did his hero Richard Spruce, the 19th century English botanist. The Professor told me that, as a six-year-old child bedridden by illness, he was given a copy of Spruce's *Notes of a Botanist on the Amazon and Andes*. Young Richard was enthralled by the tales of the explorer after whom he was to model so much of his life, as he made his way up the planet's most remote rivers, collecting plants and learning about their ethnobotanical uses along the way. Rarely does a person have the chance to fulfill such dreams through a career of exploration, collection, teaching and writing, as Richard Evans Schultes did, while at the same time inspiring a generation of botanical scientists to carry on his work around the world.

Richard Evans Schultes graduated from Harvard with an A.B. (cum laude) in 1937, having written an undergraduate thesis on Peyote (*Lophophora williamsii* (Lem.) J.M. Coult.) and its uses (Fig. 1). He undertook extensive fieldwork in Oaxaca, Mexico as a graduate student, producing a two-volume tome, *Economic Aspects of the Flora of Northeastern Oaxaca*. He went to the Amazon to study the indigenous use of curare (*Chondrodendron tomentosum* R. & P.) in 1941, supported by a National Research Council fellowship. At the outbreak of the war, he made his way from the Amazon, to Bogotá, Colombia, where he intended to enlist in the U.S. armed forces, but instead was pressed into the service of his country, collecting and studying rubber germplasm, a commodity vital to the war effort that was in desperately short supply. He remained continuously in his beloved Amazon Valley until 1953, when a Harvard administrator discovered that he had only taken out a one-year leave of absence, and it was time to return. He became Curator of the Orchid Herbarium of Oakes Ames in 1953, Curator of Economic Botany in 1958, Executive Director of the Harvard Botanical Museum in 1967, and Director of the Museum in 1970, a position he held until his retirement from Harvard in 1985. He received many honors from learned societies and governments around the world, too numerous to name in this brief piece. One of his greatest accomplishments, of which he was most proud, is his family: his wife Dorothy Crawford McNeil whom he married in 1959, and his children Richard Evans II, and the twins Neil Parker and Alexandra Ames. He would constantly speak about them to all of his students, passing along the lessons he learned about balancing family and work. He could read, write, or speak 10 languages, including two Amazonian languages, Witoto and Makuna (Fig. 2). He

Figure 2. Richard Evans Schultes and Yukuna and Tanimuka Dancers, Río Guacayá, Colombia, April 1952. Photo courtesy of R.E. Schultes.

authored numerous books and many hundreds of scientific papers, for which he was internationally acclaimed.

He would always speak warmly of his mentor, the legendary botanist Oakes Ames. Professor Ames, an economic botanist and specialist in cultivated plants and the taxonomy of the Orchid family, encouraged his student to pursue a career involving plant exploration. In a wonderfully revealing book, *Oakes Ames: Jottings of a Harvard Botanist*, edited by his daughter Pauline Ames Plimpton, there is an appendix of letters written to friends and colleagues, including the young student Richard Evans Schultes. In thinking about how his academic values and philosophy were developed, in preparation for this symposium, I read these letters and realized just how much Ames had influenced Schultes.

In an early letter to the Harvard undergraduate, Ames advised Schultes to soak up the classics and poetry, in order to help shape his writing (March 22, 1937):

My dear Mr. Schultes,

When I was a student, and that was many years ago, Charles Eliot Norton urged us read a bit of poetry every day. Indeed, one of the Boston papers got wind of this advice and printed on the first page each day a few lines from the best poets. I am sure there is much to be gained by reading a little from the great classics, and if this is done just before you yourself are about to give expression to some thought, the results are worth the interruption....And don't let a month go by without drinking in the literary charm of Shakespeare, just a page or two, now and again. This simply means that in golf, tennis, billiards, in any other game, you imitate the strokes of the masters. How much more important to enrich your vocabulary and to learn how to express yourself fluently and well in the language God has given you by reading the best English.

With my best wishes,
Oakes Ames

While thinking about his doctoral dissertation, which was heavily based on his extensive field studies in Mexico, Schultes was counseled by Ames to emphasize the importance of his applied work, and the remarkable lessons of nature, as compared to the "basic" laboratory research being carried out by many of his peers (March 4, 1940):

Dear Richard,

I desire strongly that you get away from the failings of some of those "Biological Laboratory" lads whose opinions seem to worry you. In this regard, study the philosophy of jealously. In my life-time I have read many Ph.D. theses and I once drew up some rules to cover the procedure under which some of these lads write. If you are interested here are a few of them: (1) Find some topic devoid of human interest and make it stupid. (2) Sift out every spark of human interest and write so badly regarding the residue that your ambiguity seems to imply erudition. (3) If you are capable of giving birth to a single worth-while idea, conceal it. (4) Write a cryptic summary...

And as his career progressed, Ames continued to write Schultes and encourage him to continue his work with an attitude of appreciation for humanity (March 30, 1941):

Dear Richard,

There are the big biologists and the little biologists. Most of the big biologists I have known or read about have been inferior teachers. But they have been so much imbued with the importance of humanity that they remind me of the glowing sparks which rise above the dying embers of the fire at night and seem to be carrying toward heaven a worthwhile message. The scintillating sparks rising from a kicked log on the campfire mark the difference between the stolid "standpatter" and the inspiring leader...

A few weeks later, Ames wrote about the importance of ethnobotany in a world that was globalizing, many decades before the world became aware of that phenomenon. As Western civilization spread across the world, its influence seeping into the most remote reaches of the Amazon, indigenous cultures that Schultes loved so much were faced with striking change, both in the way they lived and coping with the destruction of the environment around them. Schultes' ultimate reaction to this crisis was to sound the alarm, again decades before anyone else had realized what was happening to the tropics and its peoples (April 19, 1941):

Dear Richard,

I have presented to you the term 'treasured traditions.' When we pause to think about the

> influence of a dominating civilization it becomes clear that subjugated or submerged peoples forego many of the products and customs treasured through undated time and gradually adopt new ones. This process leads to the abandonment and gradual disappearance of many products once cherished. Unless the ethnobotanist records these while they are as yet waning, there comes a time when they are forgotten and take their place among the irrecoverable materials of human history. Hence, one of the prime duties of ethnobotany is to safeguard botanical knowledge that is on the way out, especially botanical knowledge that is integrated with human affairs…When Alexander Graham Bell tried to interest my father and his cousin F.L. Ames in the telephone, he was told that his invention had promise as a toy, but that as a practical economic contrivance it had no value whatsoever. The crucibles of progress are tended by men with vision. If you tend your crucible with an open mind, with faith in your venture, you may one day find yourself among those who have triumphed.

As Ames' passionate disciple, Schultes drank in his advice and teachings, executing his work with brilliance and precision. His command of the classics, of languages, of history, of philosophy and science provided Schultes with a perspective held by few in his day or at present. Following are a few of the many values held by the Professor, that guided the trajectory of his research and life, both in the field, classroom, and at home. There are of course many more, but I will leave the complete listing to others, perhaps as a group effort.

- The wilderness is your classroom and laboratory: As students we were advised to spend as much time in the field as possible, much in the same way that Schultes did as a student and then throughout his entire career—initially in the role of explorer, and eventually as teacher, mentor, and advisor to those seeking to conserve the Amazon region and protect its people.
- Basic biological and ethnobotanical inventory are important activities, more so when combined with hypothesis-driven, applied research questions that address real world issues: Schultes was passionate about the importance of basic inventory—of biota as well as the uses developed by people living in tropical environments. He recognized the fragile nature of both the biosphere and the ethnosphere in the regions where he worked, and encouraged us to document as much of both as possible, while at the same time shaping our work in a way that allowed it to address matters of great human concern—new sources of food, energy, medicine, fiber, and the many other necessities of both Western and indigenous culture that could be derived from nature and promote its utility and conservation.
- Great respect for indigenous people and their knowledge of the environment and its complex interrelationships: His research on the chemistry of plants used as psychoactives and medicines, the ways in which Amazonian cultures detoxified their food crops, such as cyanide-rich cassava, and their overall understanding of how ecosystems work, convinced Schultes of the sophistication of their understanding of the plants around them, and their experimentalist nature.
- Many important questions in science are best answered through a multidisciplinary lens: As the result of his broad training,

Figure 3. Photo of Richard and Dorothy Schultes from a Christmas card sent in the 1990's. Photo courtesy of R.E. Schultes.

Schultes knew what most of his colleagues at the time failed to realize—that a multidisciplinary approach to certain research questions, particularly those with an applied context, would yield much greater results than addressing the questions from a single perspective, or even a collection of single perspectives. Ames had taught Schultes as a young scientist to take off his disciplinary blinders.

- A personal commitment to lifelong learning: What fascinated me about the Professor was his humility in admitting what he did not understand or know, and that he was always ready to learn (or unlearn) something about nature.
- Working for his students, not the other way around: Schultes was exceptionally devoted to seeing his students progress in their careers, even though he had a *laissez-faire* philosophy when it came to day to day supervision. He would help his students find their way, and continue to do so, as long as he was at Harvard, and well into retirement. His students received suggestions for sources of funding, nominations for membership in prestigious learned societies, support for grant proposals, and anything else he thought could be helpful in launching or sustaining their careers.
- Gentlemanliness: In Schultes' mind, everyone was to be treated with respect, honor, and fairness in all manners of conduct and manners, or as he called it, the characteristic of "gentlemanliness."
- Devotion to family, friends, and community: This was a great and inspiring value to all of Schultes' students. He took great pride in both his academic and biological family, constantly spoke of Dorothy and the children, and felt that participation in his community was an essential part of his responsibilities as a teacher and mentor (Fig. 3).

To his students Richard Evans Schultes offered one of the greatest gifts that a teacher can give—opportunity. He was not the kind of teacher who would hold his students' hands, except to navigate some of the intricate political waters of the university. Consequently, his students were independent, took initiative, and could face successfully the many problems that fieldwork involved—from plane and boat mishaps or food shortages, to dealing with governmental visa and permit applications. These were essential survival skills for coping with life, I was later to discover. I was privileged to spend a portion of my academic life at the side of this great man, whose intelligence, courage, sense of humor, vision, and accomplishments were so inspiring. And of course, I received my series of letters from the Professor over several decades, offering guidance in my family and professional life, discussing topics of scientific concern, and reporting on his latest travels and accomplishments. His tenure as Director of the Botanical Museum was a "Golden Age" of ethnobotany and economic botany, with dozens of graduate students immersed in field studies around the world. Every week another international luminary and friend of the Professor would drop by to see him, to lecture the students, and offer their wisdom to the group.

Richard Evans Schultes left his mark on many institutions and professional associations during his career. At The New York Botanical Garden, he encouraged the establishment of the Institute of Economic Botany, initially directed by his friend and colleague Sir Ghillean T. Prance, and was an active member of its board of scientific advisors. He was a founder of the Society for Economic Botany and editor of its journal from

Figure 4. Painting of R.E. Schultes in the Nash Laboratory, with his blowgun and classroom display. Painting by Hannah Barrett; courtesy of Harvard University Herbaria.

Figure 5. The author with his professor in front of the Botanical Museum, Harvard University Graduation, June 1980. Photo courtesy of the author.

1962–1979, providing guidance to a generation of economic botanists and ethnobotanists struggling to get their data and ideas on paper and into wider circulation. He helped set up programs in economic botany at academic institutions and development projects around the world. And he was known by thousands of Harvard undergraduates for his teaching of Biology 104, the Economic Botany course that he noted was the "oldest course in the sciences at Harvard." Among the useful skills these future scientists, doctors, lawyers, and businesspeople learned was how to aim an Amazonian blowgun and shoot a curare dart into a target on the other side of the lecture hall. He showed me how to use the blowgun in the classroom, and explained how and why it worked. This particular skill was to come in quite handy during my fieldwork in the Colombian Amazon (Fig. 4).

Despite his reputation as one of the great explorers of our era, he always advised his students not to actively seek out adventures in the course of their scientific work. He once insisted to me that "if a scientist is going to the field in search of adventures he should not be there in the first place!" In my travels, which traced a small portion of his itinerary in the Colombian Amazon, I was to hear many stories about Schultes' exploits, which he rarely discussed when in Cambridge, from his former guides. It is not often that the biography of a botanist is written during his lifetime. As an example of the respect his students accorded him, Wade Davis, who did his Ph.D. under Schultes, devoted several years of his life to researching and writing *One River: Ex-*

*ploration and Discoveries in the Amazon Rain Forest*, an extraordinarily detailed and well-written book on the life and times of the Professor and his most notable student, the late Timothy Plowman.

Richard Evans Schultes, scientist, teacher, mentor, father, and friend, left an extraordinary legacy to so many of us. I am grateful to him for believing in me, for imparting some of his wisdom, and for being so patient a mentor (Fig. 5). To his family, I offer my gratitude for allowing him to share his life with us all.

## Literature Cited

Davis, W. 1997. *One River: Explorations and Discoveries in the Amazon Rain Forest.* Simon & Schuster, New York.

Ames, O. 1980. *Oakes Ames: Jottings of a Harvard Botanist.* Harvard University Faculty of Arts and Sciences, Cambridge.

# I Know Every Tree in the Forest: Reflections on the Life and Legacy of Richard Evans Schultes

Rainer W. Bussmann[1]

If anybody could be named as the father of Ethnobotany, it would be Richard E. Schultes (1915–2001). No researcher has ever done more field research, more to promote the discipline, and has encouraged more students to become ethnobotanists, with almost every senior scientist in the field acknowledging some connection to Schultes. Well known scholars like Mike Balick, Robert Bye, Wade Davis, Doel Soejarto, and Andrew Weil were all Schultesians, but Schultes also inspired generations of non-academics and always freely shared his time and advice. Schultes published almost two dozen books and almost 500 papers, collected over 30,000 herbarium numbers, often with dozens of copies, conducted continuous fieldwork in the Amazon for over 14 years, and created what can only be called the golden age of Economic Botany. How did Schultes, growing up in a poor second generation immigrant family, come to such fame? Who was this man, fluent in ten languages, who spent his entire career linked to or at Harvard, and who became the focal point of generations of useful plant researchers and plant users, and who created a by-any-account magical work setting that gave unrivaled opportunities to students, staff, and visitors? And how could it happen, that the legacy of a widely honored scholar was quickly forgotten at his home institution, and his incredible plant material and artifact collection was left to accumulate dust in a Harvard attic?

For many ethnobotanists, talking about Richard Evans Schultes (Fig. 1) is like talking about god—and how can one even attempt such a thing, especially when one has never had the privilege to meet the man? This was a challenge I faced when I was asked to contribute a lecture on Richard Schultes to the "Botanists in New England" Symposium of the Economic Botany and History Sections of the Botanical Society of America in 2010. My initial strategy, and perhaps the only way to appreciate Schultes' thinking, was to simply head out into the field, somewhere in the upper Amazon, to collect plants and think about how to meet this challenge (Fig. 2).

Apart from *One River*, Wade Davis' biography of Schultes (Davis, 1996), astonishingly little has been written about the founder of modern ethnobotany. Even the most complete bibliography in the obituary by Prance (2001) is far from presenting a complete picture of Richard Schultes' extensive writing. How could it happen that the "father of ethnobotany" is nowadays often unknown to students, and maybe only vaguely known to colleagues, despite his reputation as the man who "discovered" almost every single psychoactive species known in the New World?

For example take myself. When starting my ethnobotanical work, a friend asked me "So, do you know Richard Scholtees?" (At least that is how his name sounded to me, and the mispronunciation certainly did not help. I simply had no clue who that might have been.) So, "Richard who?" was my understandable answer. Schultes himself would most certainly have flinched at the mangling of his name. He was very well known for insisting on the correct spelling and pronunciation of Latin, Greek, and any other language.

Who was this man, who collected over 25,000 botanical specimens, wrote two dozen books and almost 500 scientific papers?

[1] William L. Brown Center, Missouri Botanical Garden, P.O. Box 299, St. Louis, MO 63166-0299, USA. Email: rainer.bussmann@mobot.org.

Figure 1. Richard Schultes in the field. Photo courtesy of the Schultes family.

Schultes, pronounced like the "u" like "o" in "who", and the "e" like "e" in "the," had an unlikely background for someone who was to become the world's most famous ethnobotanist. Born in 1915 to a poor working class family in East Boston with German and English roots, Schultes excelled at school, especially in the Classics, and ultimately earned a stipend to attend Harvard Medical School in 1934. Looking to fulfill his pre-med science requirements, he happened to walk into Oakes Ames' Economic Botany class, a fateful event that determined the course of the rest of his life.

When asked to review an ethnobotanical study for a class assignment, Schultes, hard pressed for time, grabbed the thinnest volume on the book desk, *Mescal: The Divine Plant and Its Psychoactive Effects* by Heinrich Kluever, and literally got hooked. The very next class Schultes asked Ames if he could write his thesis about the plant, and from this point his research life inevitably took on a different character.

In 1936, Schultes, together with colleague Weston LeBarre, piled into an old Studebaker and headed west from Massachusetts to Oklahoma to study peyote use amongst the Kiowa. It is hard to imagine how difficult it must have been in those days to make such a trip. Over endless days on rutted roads the car broke down constantly. What also made the trip unusual was to have a botanist going into the field to talk to people. Botanists traditionally ventured out to collect plants. "Collecting stories" was more the task of anthropologists or linguists - "soft scientists" - as many molecular oriented biologists still believe. On this trip, a botanist and an anthropologist were working together! For Schultes, the experience of being in a strange setting under the full effects of the *Lophophora* compounds, must have quite profound. Photographs from the journey, nevertheless, always show a perfectly groomed Schultes, while LeBarre definitely shows the effects of all-night peyote ceremonies. This was to become a trademark of Schultes: always conservatively and correctly dressed, rather more like a Victorian professor than somebody who during the 1960s would lecture on mind-altering substances. Only one year after his first trip to the Kiowa, the young Schultes was found testifying to congress in support of the legalization of peyote in traditional ceremonies. Quite a stand for a young, unknown botanist.

Figure 2. Rainer Bussmann in the field, Alto Marañon, Peru, 2010, musing about R.E. Schultes. Photo courtesy of the author.

Despite his conservative appearance (Schultes is rumored to have always voted in presidential elections for Elizabeth II as a write-in candidate and always lectured in coat, tie, flannel trousers and a white lab coat), Schultes was a great advocate for personal choice. He abhorred government interference in one's personal life and decisions, and despised the Kennedy clan, whose patriarch made his fortune in bootlegging during prohibition, when the brewery where Schultes' father worked was forced to shut down.

After finishing his undergraduate thesis on the Kiowa's plant use, young Schultes immediately embarked on another project, the identification of "Teonanacatl" and "Ololiqui," two plants sacred to the Aztecs that had defied attempts to scientifically identify them. The task led Schultes, by any kind of transport imaginable, deep into the then most remote corners of Mexico (Fig. 3). In a remarkably short time, the young botanist managed to solve a mystery that had confounded colleagues for centuries. Teonanacatl turned out to be a group of mushrooms, mostly *Psilocybe* spp. and *Paneolus* spp., while Ololiqui was identified as *Turbinia corymbosa* (L.) Raf., a morning glory. Both contained essentially the same psychoactive compounds chemically similar to LSD, and it would take more than 60 years to prove that the psilocybin in the seeds of *Turbinia*, identical to the compound

Figure 3. Oaxaca, 1938. Photo courtesy of the Schultes family.

found in "magic mushrooms," is in fact produced by a fungus that frequently infects the plant, rather than by the vine itself. The re-discovery of the identity of these two Mexican hallucinogens by Schultes earned him a Ph.D. in 1941. His work would later become the foundation of the psychedelic era, when colleagues, like Timothy Leary, dug up Schultes' papers on the subject, started to experiment with the plants, and made the properties of the plants known to the public. Schultes himself would always criticize the recreational use of plants and their psychoactive compounds. This, however, did not keep him from traveling the states to defend students accused of smoking marijuana. His argument was simply that *Cannabis indica* (the prohibited species) was in a ground up stage not distinguishable from *Cannabis sativa*, which was still widely grown for fibers and perfectly legal at that time. This effort certainly added to Schultes' fame among students and colleagues.

After finishing his Ph.D. in 1941, Dr. Schultes faced two choices: Either start teaching at a small college, or accept a government assignment to head to the Amazon to collect data on arrow poisons. Oakes Ames had offered the following advice:

"If I were a young man beginning my career all over again, I should try through intensive research in economic botany and ethnobotany to bring more light into the intellectual realm and to take my place, not in a laboratory cubicle, but in the world..." (Oakes Ames to Schultes 1941; Ames 1980).

Schultes chose to head for the Amazon, not knowing that this would keep him in the Columbian Amazon almost without interruption until 1953 (Figs. 4 and 5).

America's entry into WWII found Schultes deep in the tropical forest. Returning to Bogotá to volunteer, he was rapidly incorporated into the war effort. Natural rubber supplies came at that time, and still come, mainly from plantations in SE Asia. These were occupied by the Japanese in 1942. Although the U.S. petrochemical industry started a synthetic rubber program, and the USDA planted rubber yielding plants, especially dandelions (*Taraxacum officinale*), in almost every state, rubber from *Hevea* spp. was simply irreplaceable. Planting the species in the new world (e.g., in the humongous *Hevea* plantations at "Fordlandia" in Brazil) had failed because of rampant leaf blight infections. To remedy this situation, the U.S. government established the Rubber Reserve Company, to collect all recyclable scraps of rubber, and find new sources in nature. Needed most were blight resistant varieties, as well as a way to again extract rubber from natural populations found in the Amazon basin. Both tasks were part of the assignment given to Schultes that would keep him busy for most of the next 14 years. During this time, Schultes traveled an area the size of Belgium, mostly on foot and dugout canoe, mapping thousands of individual rubber trees and providing seeds for the establishment of

Figure 4. Richard E. Schultes in Columbia, between 1946 and 1952. Photo courtesy of the Schultes family.

Figure 5. Macarena, Columbia, 1951. Photo courtesy of the Schultes family.

blight resistant trial plantations in Colombia and Central America. At the same time, he managed to collect almost 25,000 herbarium numbers, and documented the uses of 2,000 plant species among two dozen different tribes. The attitude that made his task possible was certainly his willingness to respect everybody, treat them fairly, and honor the value of local knowledge (Fig. 6). This work, in particular during the first half of the 1950s, would be the basis for over 400 papers and two dozen books.

In 1953, the government suddenly reversed its position on natural rubber. Some bureaucrat decided that, with the end of the war and the advance of synthetic rubber, the natural rubber program was no longer needed. Schultes and colleagues had to wrap up work and leave. The trial plantations were ultimately cut down, the collected material was forgotten. This would prove a rather shortsighted decision, as we will see later.

Schultes returned to Harvard and in 1953 became Curator of Oaks Ames' Orchid Herbarium. As a result, he had to concentrate all his efforts on orchids, which left virtually no time for ethnobotanical work. In 1958, Schultes was finally given the position of Curator of Economic Botany. Despite his tireless work, however, it took another decade before he was named Executive Director of the Harvard Botanical Museum in 1967, and Director in 1970. During his tenure at Harvard, Schultes worked steadfastly on the ethnobotany of the North-

Figure 6. Yayacopi, Rio Apaporis, February 1952. Photo courtesy of the Schultes family.

western Amazon, describing a plethora of new species from his collections, and publishing the results of his long term Amazonian studies.

Although Schultes for many years taught one of the most popular courses at Harvard, and had found his real vocation in teaching and advising students, he only became a professor in 1973. Many of his students went on to become well known ethnobotanists and medicinal practitioners, including Tim Plowman (deceased), Mike Balick (New York Botanical Garden), Bob Bye (Universidad Nacional Autonioma de Mexico), Djaja Soejarto (Chicago Field Museum), Jim Zarucchi (Missouri Botanical Garden), Andrew Weil (Weil Lifestyle), Mark Plotkin (Amazon Conservation Team), and Wade Davis (Explorer in Residence at National Geographic Society).

Schultes encouraged students to follow their dreams, rather than working on thesis subjects of the kind described by Oakes Ames in a letter to Schultes in 1940:

> ...cover the procedure under which some of these lads write.... 1. Find some topic devoid of human interest. 2. Sift out every spark of human interest and write so badly....that your ambiguity seems to imply erudition. 3. If you are capable of giving birth to a single worth-while idea, conceal. 4. Write a cryptic summary. (Ames 1980)

Among his many accomplishments, Schultes published the *Harvard Botanical Leaflets* for decades, was editor of the journal *Economic Botany* for 15 years, wrote hundreds of book reviews for the journal, and was a leading figure not only in the establishment of the Society for Economic Botany, but also in the creation of the International Society for Ethnopharmacology. Simultaneous to his academic work, Schultes started to advocate the need for conservation in areas like the Amazon, as well as the preservation of traditional knowledge associated with plants found there. During his career, Schultes received countless honors, including Distinguished Economic Botanist (1979), the Tyler Ecology Prize (1987), gold medals of both the Linnaean Society and the World Wildlife Fund (1992), election as a Global 500 Forum laureate (1994), and the highest order of Columbia given to both civilians and military.

However, despite his publication and teaching record, and despite all national and international recognition Schultes received, Harvard did not deem it necessary to maintain his legacy. As soon as he retired in 1985, his journal was closed, his famous lab dismantled, and his huge collection of ethnobotanical material placed in an attic, where it remains, largely uncurated, to this day.

One of the plants Schultes' students worked on intensively was coca (*Erythroxylon novogranatense* (Morris) Hieron and *E. coca* Lam.), the sacred leaf of the Andes and today the notorious source of cocaine. Schultes' students noted decades ago that the traditional use of the plant is in fact harmless, contributes to a balanced nutrition, and had nothing whatsoever to do with drug use and production. The plant has nevertheless now become the central focus of the so called "war on drugs." Rather than addressing the social problems that foster drug use in the "developed" world, including the US, the U.S. government has spent billions of dollars in recent decades to wage chemical war, fumigating millions of acres of coca fields, especially in Columbia. This effort very much reminds me of the use of Agent Orange in the Vietnam war, and coincidentally the herbicides used in the war on drugs are produced by the same companies that supplied defoliants to the military during the 1960s and 1970s. Unfortunately, the coca eradication program destroys not only coca, but all other crops, like coffee and corn, and has detrimental effects of the health of the local population. Ultimately, the chemicals affect the whole ecosystem, and with it the traditional knowledge Schultes took great care in documenting.

This brings us back to Schultes' original rubber work in the Amazon. Despite the advances in synthetic rubber production, natural rubber remains the material of choice for highest quality products and applications, from gaskets, space and military applications, to surgical gloves and condoms. Even radial tires cannot be produced without natural rubber, and 95% of the production is still coming from plantations in SE Asia. Unfortunately, due to faster transport possibilities in a global economy, leaf blight has finally reached that region, and while outbreaks have been small, the danger exists that the supply of one of the most important commercial and strategically important natural products could be disrupted by a large scale outbreak of blight, whether "natural" or caused by an act of bioterrorism.

Sadly enough, the blight-resistant clones Schultes and colleagues collected have been destroyed, and his priceless collection is slowly turning to dust. Once lost, the material will be hard to replace, because botanists who can correctly identify the material are themselves critically endangered; most Botany programs in the U.S. have been replaced by "more useful" molecular science. The regions where Schultes collected, even if collection permits were issued (which is very unlikely, given that the U.S. still has not ratified the Convention of Biological Diversity), are coincidentally exactly the areas that we have systematically fumigated as part of our ill-advised strategy to eradicate coca, and little of the natural vegetation remains.

German colleagues have already reverted to what *The Economist*, oblivious to the efforts of the Rubber Reserve Program, named one of "The 50 best inventions 2009:

> "DANDELION Rubber: A fast-spreading fungus is ravaging the world's rubber trees. But thanks to researchers at Germany's Fraunhofer Institute for Molecular Biology and Applied Ecology, there's now an alternative: dandelions. Scientists have long known that the weed's sap contains latex, but it's difficult to harvest because dandelion ooze polymerizes—goes gummy—when it hits the air. The Fraunhofer team overcame that sticky problem by switching off a key enzyme. The new, improved dandelion produces 500% more usable latex than the old weed does."

Welcome to the past....

Richard Schultes died 10 years ago, and it is high time to honor his legacy and make sure that his groundbreaking work is no longer neglected.

## Literature Cited


Ames, O. 1980. *Oakes Ames: Jottings of a Harvard Botanist.* Botanical Museum of Harvard University, Cambridge, MA.

Davis, W. 1996. *One River: Explorations and Discoveries in the Amazon Rainforest.* Touchstone, New York.

Prance, G. 2001. Richard Evans Schultes (12 January 1915 – 10 April 2001): A Tribute. *Econ Bot* 55(3):347–362.

# Traditional Knowledge in a Changing World: New Insights from the Chácobo in Bolivia

Rainer W. Bussmann[1] and Narel Y. Paniagua Zambrana[2]

The Chácobo are a Panoan speaking tribe of about 500 members in Beni, Bolivia. Originally nomadic, the Chácabo were relocated to their current main location in the 1960s, after American missionaries convinced the Bolivian government to set up a homeland for the tribe, and migration ceased. The first large ethnobotanical study on the tribe was conducted by Brian Boom in the 1980s. During the last decade, the status of indigenous groups in Bolivia has markedly improved, and the Chácobo now have legal title to a territory of about 450,000 hectares. Our present work represents the start of a follow up ethnobotanical inventory of the tribe. The departure of permanent missionaries, access to markets, with an all-weather road constructed in the last few years, as well as a larger territory, seem to have already markedly changed plant use and knowledge patterns and the lifestyle of the population. All children now learn Chácobo as their first language in school, the use of important plant species centers more on income generation, and the more sedentary lifestyle of the tribe has been replaced by migration patterns, based on seasonal availability of resources.

## Introduction

Bolivia has a rich diversity of indigenous cultures, with at least thirty tribes among eleven language groups. However, there are few detailed studies on the use of plants and resources by the people of the area, and their knowledge of forest management is still being ignored by the authorities. The Summer Institute of Linguistics (SIL) formerly had a strong impact in changing the culture of many tribes in Bolivia, including the Chácobo. The SIL worked with Chácobo communities from 1953 to 1980, when it was replaced by a Swiss Protestant mission established in Riberalta, leading to a profound change of lifestyle, and a permanent process of acculturation. Brian Boom (Boom 1987) led the first ethnobotanical study of Chácobo between 1983–1984, documenting their knowledge after almost 30 years into this cultural change.

The Chácobo belong to the Panoan linguistic group that includes about twelve tribes (Chácobo, Pacahuara, Matis, Matses, Yaminahua, and others).

At the end of the 1890s, the Chácobo lived as semi-nomadic hunters and cassava and corn cultivators, probably in two groups, one with six families and one with four, in northwestern Bolivia, roughly between Lake Roguagnado and the river Mamcre, south of their current territory. During the rubber boom in the early 1900s, they were forced by more aggressive tribes to move north, where they were threatened by rubber tappers, who also brought disease and epidemics to the tribe. The Chácobo managed to avoid most of the outside influences however, while other tribes in the region were hunted like animals to be enslaved to work in rubber stations. The Chácobo had their first permanent contact with the outside world only in 1953 with people from the New

[1] William L. Brown Center, Missouri Botanical Garden, P.O. Box 299, St. Louis, MO 63166-0299, USA. Email: rainer.bussmann@mobot.org.

[2] Herbario Nacional de Bolivia, Universidad Mayor de San Andrés, Casilla 10077 – Correo Central, La Paz, Bolivia. Email: nyaroslava@yahoo.es.

Figure 1. Contemporary Alto Ivon, Beni, Bolivia. Photo courtesy of the authors.

Tribes Missions, and in 1954 the Bolivian government established an agency about 15 km from the current location of Puerto Limones. The missionary linguist Gilbert Prost arrived in 1955 under the auspices of the Summer Institute of Linguistics (SIL). According to Prost there were four Chácobo groups living between the Benicito and Yata rivers, with about 200 people (Boom 1987).

Prost and his wife continued to live among the Chácobo until 1980. In addition to translating the New Testament into Chácobo, they made some observations on cultural and linguistic practices. In 1964, Prost managed to buy a territory in the north of the Chácobo's ancestral lands, forming the community of Alto Ivon (Fig. 1), and most of the remaining population moved there. In 1965, the Bolivian government finally assigned 43,000 hectares of land to the Chácobo, although this area was less than 10% of their original territory (Fig. 2). The influence of Prost caused profound cultural change among the Chácobo, including the abandonment of traditional costume and dance in 1969 (Fig. 3).

Currently the population of the Chácobo community numbers about 500 people, with Alto Ivon as the largest settlement and Tokyo, Motacuzal, Siete Almendras, and other smaller communities along the Yata River. The current territory of the tribe encompasses 450,000 hectars (Fig. 4), and is roughly equivalent to the original extent of the tribe's ancestral lands. The community of Alto Ivon, the center of the Chácobo territory, is located about 112 km south of Riberalta, along the river Ivon, a tributary of the

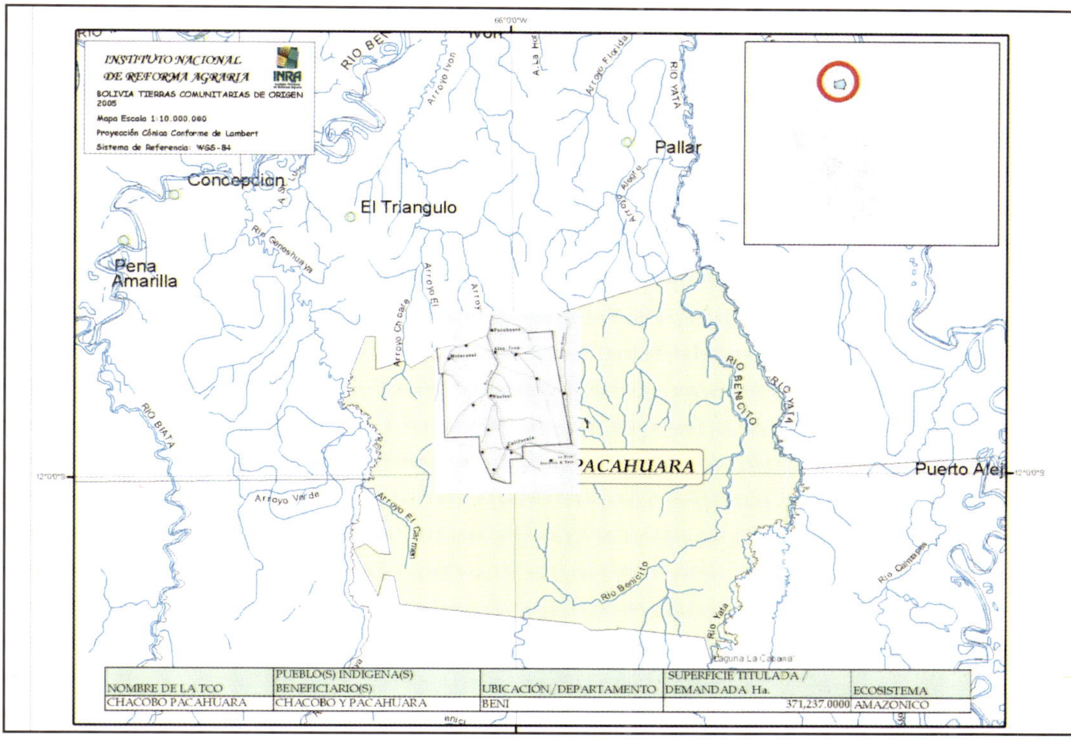


Figure 2. Chácobo territory 2010, with 1986 territory overlaid in grey. Map courtesy of CIRABO.

Beni. The elevation is about 200 m, and can be classified as Amazon rainforest. Rubber trees (*Hevea brasiliensis*) and Brazilnut (*Bertholletia excelsa*) are abundant. The average annual temperature is 26.8°C, with an average annual rainfall of 1.56 mm, based on observations in Riberalta. A distinct dry season lasts from June to November.

Formerly the Chácobo were led by a Cacique. Today there are two indigenous organizations: The Capitanía Mayor Chácobo, closely linked to the evangelists, and the Chácobo-Pacahuara Association, recognized by the Central Indígena de la Región Amazónica de Bolivia (CIRABO), supported by the Central de Pueblos Indigenas del Beni (CPIB) and the Confederacion de Pueblos Indigenas de Bolivia (CIDOB).

## Methodology

In October 2011, the authors visited the Chácobo to conduct semi-structured interviews about the use of Arecaceae in the community. After obtaining prior informed consent from CIRABO as well as the community members, 89 randomly selected Chácobo (with equal gender and age distribution) were interviewd. In the process, we observed profund changes in knowledge as compared to what was reported by Boom (1987). The presented contribution gives a first outline of these changes.

## Changes in Chacobo life 1985–2011

Certainly the most striking examples in Chácobo life initially observed by an outside visi-

Figure 3. Chácobo family 1986 in mission style dress (Boom 1987). Photo courtesy of NYBG Press.

tor are changes in fashion and use of traditional tools. While missionaries introduced a rather conservative clothing style, nowadays the Chácobo population simply uses regular "modern" clothing as available in the markets of Riberalta (Fig. 5).

The changes in traditional implements are more subtle. Every house still has one or more mortars to de-husk rice, which is the most frequently used implement in daily life and has not changed since Boom's observations. Large clay pans for roasting Manihot flour (jibé) are also still found in most houses (Fig. 6). In the 1980's, cassava (*Manihot esculenta*, Euphorbiaceae) was clearly the most important food for Chácobo, who planted seven varieties covering almost 62% of all cultivated land in the community. Cassava was harvested from May. Maize (*Zea mays*, planted on 18% of the land) and upland rice (*Oryza sativa*, planted on 7% of the land), both Poaceae, were planted alternately with cassava (Boom 1987). By 2010, rice has become the staple food, leaving cassava and corn with a more secondary role, clearly reflecting a change in the Chácobo diet, and also showing the influence of the market, because a large part of the rice crop is sold. Thus, most of the roasting pans are in a rather dire state of repair, due to infre-

Figure 4. Chácobo territory 1986 (adapted from Boom 1987). Map courtesy of NYBG Press.

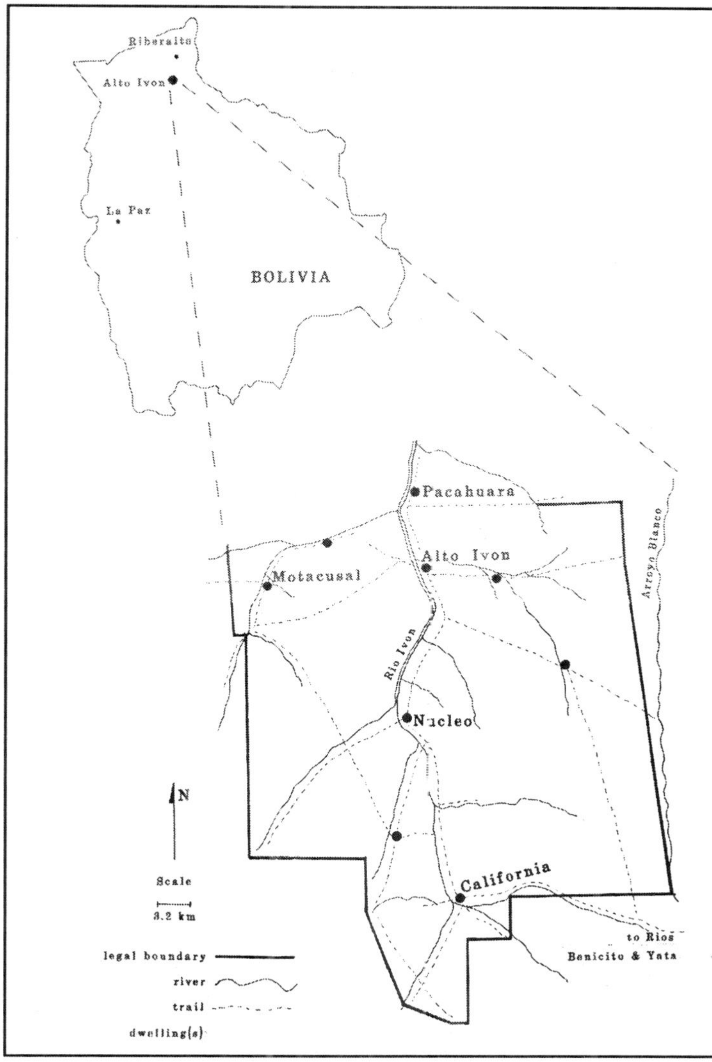


quent use. Only in the more remote and smaller Chácobo villages could we observe women in progress of making jibé. How to harvest and dry the rice has also changed. Now the whole plant is harvested and stored in warehouses, without drying the grain in the field.

Another important crop was banana or plantain and banana (*Musa* x *paradisiaca*, Musaceae), planted in gardens and clearings, without special place on 12% of the land, but with obvious importance in the food they ate. Boom found eight local varieties of banana (Boom 1987) (Fig. 7). By 2010, bananas had completely lost importance, and only the standard Cavendish variety was planted as a secondary crop around the houses. All the traditional varieties had disappeared, indicating a profound loss of traditional plant genetic resources.

Boom mentions that 27 edible plants were grown in small gardens laid out near homes,

Figure 5. Chácobo family dress in 2011. Photos courtesy of the authors.

the most common species being mango (*Mangifera indica*, Anacardiaceae), avocado (*Persea americana*, Lauraceae), papaya (*Carica papaya*, Caricaceae), and lemon (*Citrus aurantifolia*, Rutaceae) (Boom 1987). In our work done in 2010, we found only mango and orange (*Citrus sinensis*, Rutaceae) in these home gardens.

Shrubs such as *Psidium* sp., *Myrica* sp., and *Eugenia* sp., all Myrtaceae, were also common during the 80's. Another plant commonly grown in home-gardens was *Bixa orellana* (Bixaceae), used for coloring food, as well as melon (*Citrullus vulgaris*, Cucurbitaceae) and sugarcane (*Saccharum officinarum*, Poaceae). In 2010, we found few instances of these species,

Figure 6. Chácobo food production in 1986 (Boom 1987) and 2011. B&W photos courtesy of NYBG Press. Color photos courtesy of the authors.

while pineapple (*Ananas comosus*, Bromeliaceae), bean (*Vigna unguculata*, Fabaceae), yams (*Dioscorea trifida*, Dioscoreaceae), waylusa (*Xanthosoma* sp., Araceae), sweet potato (*Ipomoea batatus*, Convolvulaceae), and squash (*Lagenaria siceraria*, Cucurbitaceae) are still an important component of local diet.

In the 1980's, the Chácobo also often sowed Huanimá (*Bactris gasipaes*, Arecaceae) in abandoned clearings to collect palm fruits. By 2010, the palm was only found rarely around the villages, and was no longer planted.

A great diversity of forest plants were used for medicinal purposes by the Chácobo. Boom's informants mentioned 174 species, but often did not indicate exactly how they were using each of them. They most often described the preparaton of remedies by boiling the leaves, bark, or fruits to cure diseases. Sometimes other plant parts (leaf, bark, latex, resins, fruits) were applied directly to the body region that was affected, particularly in cases when using latex or other plant extracts. However, Boom did not find a true healer among the Chácobo who would have been able to corroborate or supplement the information obtained from his informants (Boom 1987). The knowledge of medicinal plants was particularly alive among older informants interviewed, an indication that traditional medici-

Figure 7. Variety of bananas eaten by the Chácobo in 1986 (Boom 1987). Photo courtesy of NYBG Press.

nal knowledge was already in the process of being lost in the 1980s.

Similarly, the use of plant poisons still had importance when Boom visited the Chácobo. The latex of *Odontadenia laxiflora* (Apocynaceae) was applied to arrowheads to catch fish in deep water. *Derris floribunda* (Fabaceae) was only cultivated for fishing, while *Derris amazonica*, *Serjania lethalis*, and *S. pyramidalis* (Sapindaceae) were collected only in the forest. Stems of these species were macerated and applied to shallow water during the dry season to suffocate the fish, which were collected when they came to the surface. In 2010, such fish poisons had essentially disappeared.

Bows and arrows were originally the most important fishing and hunting implements. The Chácobo most frequently used the hard wood of palms like *Astrocaryum aculeatum* and *Bactris gasipaes* (Arecaceae) to manufacture bows and arrows. The handles of the arrows were made from the flowering stems of wild cane (*Gynerium sagittatum*, Poaceae), and when this was not available the stems of *Geonoma maxima* var. *chelidonura* (Arecaceae) or *Ischnosiphon arouma* (Marantaceae) (Boom 1987) were used. Boom identified five very different types of arrows designed to hunt different types of animals: Tepi and quërëquë (for ground birds), tohua quëspini and paca (for large animals), and bicobi (for fish). In 2010, the Chácobo were only using .22 caliber rifles and 20 gauge shotguns for hunting, and only bicobi was still used for fishing. Other types of arrows had disappeared completely (Fig. 8).

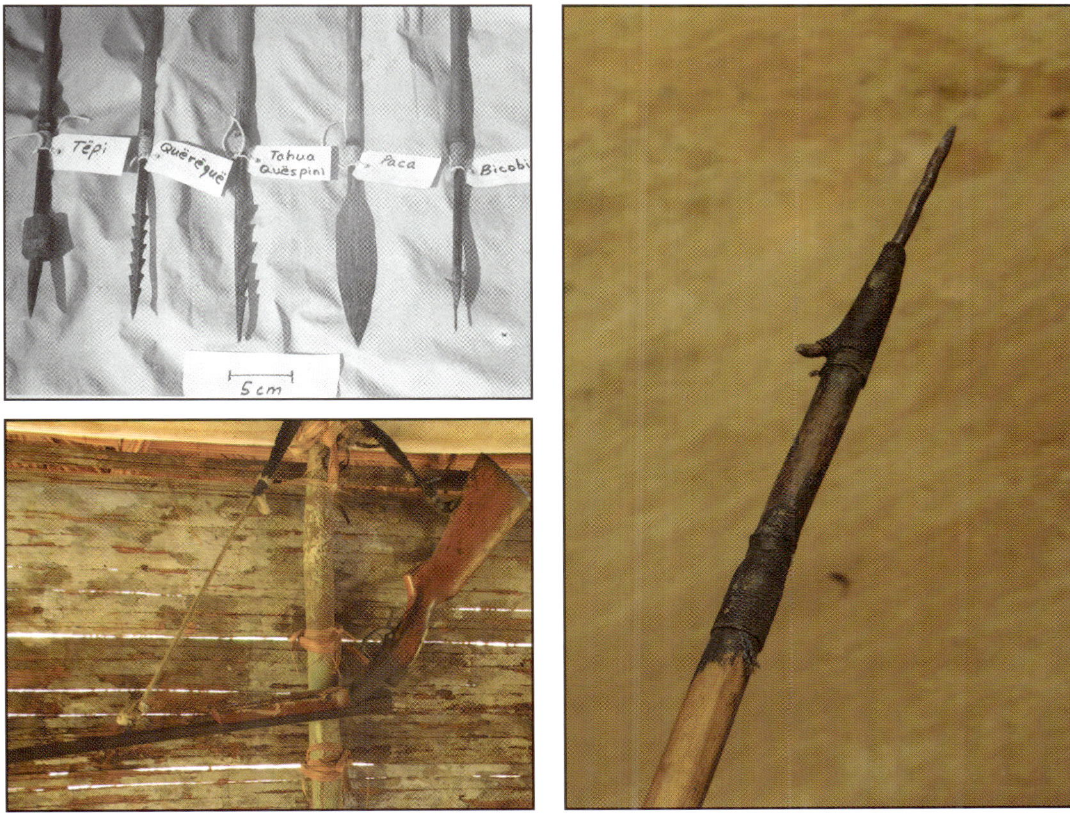


Figure 8. Chácobo arrows 1986 (Boom 1987) and hunting implements from 2011: a shotgun and bicobi, the only arrow form left in 2010. B&W photo courtesy of NYBG Press. Color photos courtesy of the authors.

Many species were used to make baskets (Fig. 9). The ropes for large burden baskets (cacano) were made from the inner bark of *Couratari guianensis* (Lecythidaceae), and *Uncaria guianensis* (Rubiaceae) is used for the frame. The strap that is placed on the forehead of the person is made from the inner bark of *Ficus obtusifolia* (Moraceae). The poropachi, a light cargo basket, is woven from young leaves of *Attalea phalerata* (Arecaceae). This has not changed since Boom (1987), and the same species are still used to make baskets. However, only a few older women in the communities still have the skills to weave baskets, and modern implements like backpacks are clearly replacing traditional materials. The cacano nowadays mostly serves as a storage basket, or for chickens to lay their eggs in.

One of the most profound changes in Chácobo life seems to be a return to former migration patterns, nowadays mostly linked to commerce and income generation.

Originally Brazil nuts (Castaña, *Bertholletia excelsa*, Lecythidaceae) were an important source of fats in the Chácobo's diet. People still eat the nuts raw or roasted, but most of the population collects Castañas only for sale. In the 1980s, Brazil nuts sold for 2,500 Bolivianos per kilo (U.S. $2.50 in 1983) in Riberalta. Jihui (rubber, *Hevea brasiliensis*, Euphorbiace-

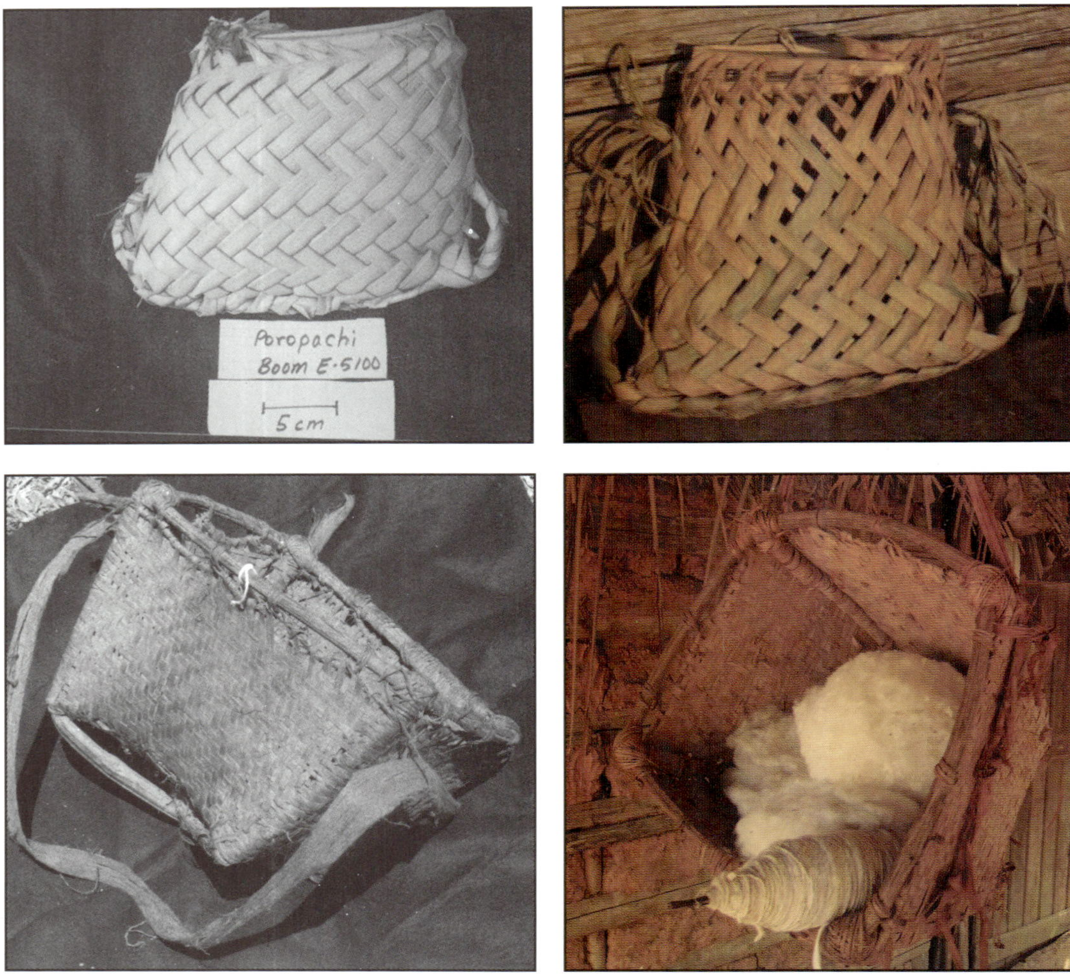


Figure 9. Chácobo basketry in 1986 (Boom 1987) and 2011. B&W photos courtesy of NYBG Press. Color photos courtesy of the authors.

ae) was then sold for 1,400 bolivianos per kilo (U.S. $1.40 in 1983). At that time, the rubber business was the most important source of income for the Chácobo. By 2010, the marketing of these products had changed dramatically. Rubber has no importance whatsoever as a source of income, and is only used traditionally to manufacture some household utensils. The harvest of Brazil nuts, meanwhile, now represents the main source of income for many Chácobo. A product that has recently become more economically important is the palm heart from Acai (*Euterpe precatoria*, Arecaceae), which is collected by walking increasingly longer distances and sold to contractors who enter the communities to purchase large quantities delivered to the canning factories in Guayaramerín and Riberalta.

The production of oil from the seeds of Brazil nuts was reported by Boom (1987), but has completely ceased nowadays. The chestnut harvest takes place from January and March,

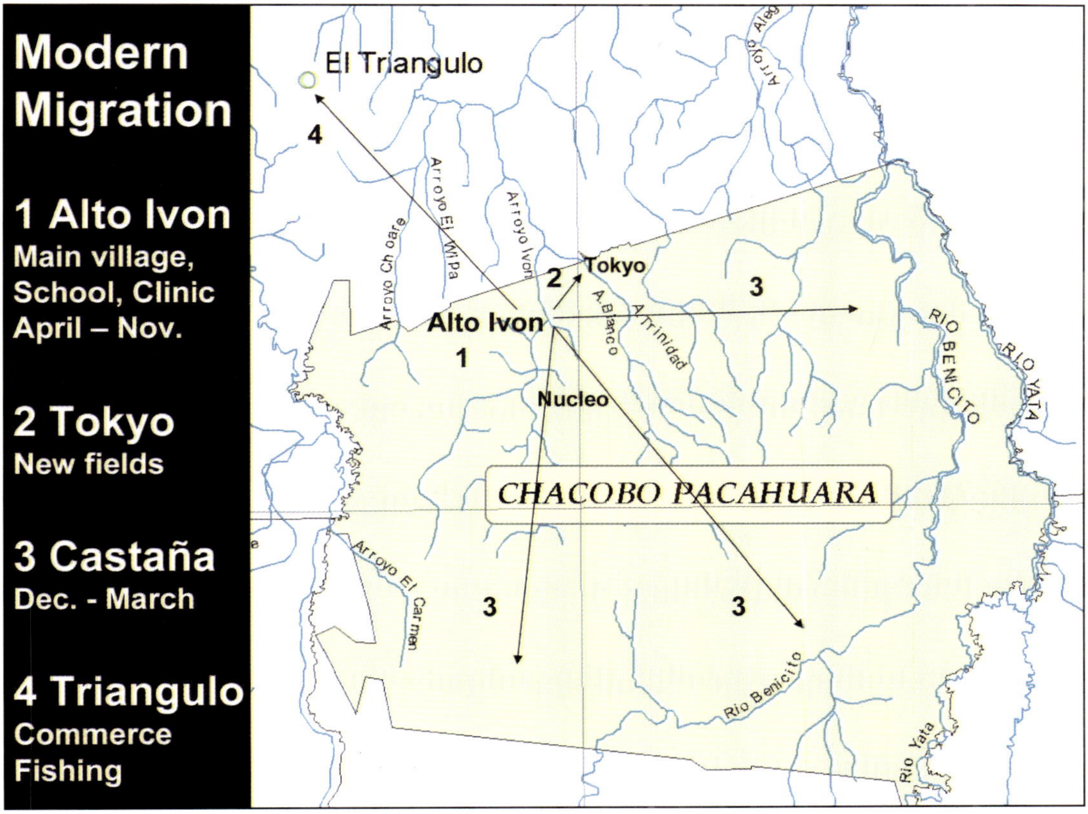


Figure 10. Modern day Chácobo migration patterns. Map courtesy of CIRABO.

and during that time now almost the entire Chácobo population migrates to the South of the territory where the largest concentration of *Bertholletia* is encountered. During the rest of the year Alto Ivon remains the main population center. However, many Chácobo have "second" homes in Tokyo, where most of the fields are located at present, or in Triangulo, closer to their main fishing sources, and conveniently located at the road to Riberalta, leading effectively to a semi-nomadic lifestyle (Fig. 10).

## LITERATURE CITED

Boom, B.M. 1987. Ethnobotany of the Chácobo Indians, Beni, Bolivia. *Adv Econ Bot* 4:1–68.

# Passage through Mexico—The Influence of Richard Evans Schultes on Mexican Economic Botany

Robert Bye[1]

Richard Evans Schultes dedicated his life to studying the interrelationships between plants and people as well as sharing his enthusiasm with members of academic and public circles. His botanical formation was grounded in the Harvard University's tradition of taxonomy and "useful" plants. After completing original field work on peyote in Oklahoma (USA), his interdisciplinary vision shifted to Mexico. At these southern latitudes, he discovered more members of this mind-altering ethnobotanical complex and pioneered interdisciplinary studies on mushrooms and ololiuqui. After completing his doctoral thesis on the useful plants of northeastern Oaxaca in 1941, he moved on to northwestern South America, where his focus resided for the remainder of his life, and where he built his international eminence. Northeastern Oaxaca was the area in Mexico with which RES became best acquainted. His introduction to the area by the Austrian B. P. Reko allowed him to survey the economic plants of the mountains of the Sierra Madre of Oaxaca; later, he presented the area, the people and the sacred mushrooms to ethnomycologists, such as Roger Heim and Gordon Wasson. His broad interest as "the last Victorian explorer" followed the paths by European botanists of the previous century. Other plants caught his attention and he began to publish a series (Plantae Mexicanae) dedicated to expanding the taxonomic, floristic, and economic botany of this region. In the late 1950s, he ended the publication of his original Mexican research with articles on orchids and the Oaxacan spice, *Quararibea funebris*. Although northwestern South America dominated his field work, he maintained contact with the second center of New World sacred narcotics through collaboration with researchers conducting field and laboratory investigation in Mexico. Even though the public does not associate RES with Mexican ethnobotany, the intellectual foundation and the implementation of Mexico's contemporary programs in sacred plants, medicinal plants, interpretation of plants found historical sources, and maize conservation are strongly influenced by RES's Mexican passage.

Richard Evans Schultes' evolution from his peyote studies in Oklahoma to the diverse discoveries of medicinal and hallucinogenic plants of the Amazon occurred in Mexico. During this process, he not only integrated peyote into the complicated cultural, biological, and historical context, but also propelled ethnobotany and economic botany to resolve concrete problems (Herrera et al. 1998).

His initial training and life-long commitment was rooted in botanical sciences at Harvard University which emulated the development of the Royal Botanic Gardens Kew. In 1858, Asa Gray, the dean of American botany working at Harvard, wrote to William Hooker that he established the 'Miscellaneous collections' in his Cambridge herbarium "...in humble imitation [of the Museum of vegetable products] of Kew ...". Later in 1888, George Lincoln was hired to assist in teaching plant physiology and curate the museum so that it would "...function as illustrative material for the teacher and as a reference collection [for] the specialists...;" it evolved into the Botanical Museum of which Goodale was its first director (Dupree 1959). Dr. Oakes Ames succeeded as curator and director and focused his research and teaching on orchid taxonomy and the theme: Plants

[1] Jardín Botánico del Instituto de Biología, Universidad Nacional Autónoma de México, 04510 México, DF, México.

and Human Affairs. Ames so impressed the young Schultes that he became the Bostonian student's mentor. When RES was appointed director of the Botanical Museum in 1968, he carried on with the course (which was taught then by the Botanical Museum director who succeeded Ames, Paul C. Mangelsdorf) until his retirement in 1985.

RES fostered a special interest in the Orchidaceae. He pursued the taxonomy of this tropical family and delved into its biogeography and history. His first formal museum position was that of Curator of Orchid Herbarium (AMES) between 1954 and 1958 (now part of the Harvard University Herbaria or HUH). However, this taxonomic field was shadowed by areas of parallel importance—economic botany and ethnobotany.

RES's legacy to ethnobotany is multifaceted. He always insisted on the respect for collaborators and colleagues in both the academic world as well as in the field around the world. He acknowledged and built upon the historical background of different disciplines that interfaced within the sphere of plants and human affairs. He maintained a firm taxonomic basis and the solid underpinning of regional floras and the biogeography of their elements. Emphasis was placed on indigenous names, classification, values, and traditional uses of the plants. One must document the changes and advancement in plant-human interactions and their potential for the advancement of human societies. He distinguished economic botany from ethnobotany in the sense that the latter pertained to "primitive" people; the distinction of the latter by such a label was part of his definition that provoked much discussion between him and his students during the 1960s and later.

RES' career is marked by 355 publications that deal with various aspects of ethnobotany (Prance 2001). His printed legacy to Mexican ethnobotany is found in only 27 articles (7.6% of his total output), of which 13 were published between 1940 and 1941. He contemplated two series of articles: *Plantae Mexicanae* and *Contribuciones al conocimientò de la botánica de Oaxaca.* Unfortunately, these were truncated with only ten and three installments, respectively. The original field data for these papers were derived from his 5 months of field work in northeastern Oaxaca during 1938 and 1939. The comparative study of his findings with contemporary and historical sources was presented in 1941 as his doctoral thesis, *Economic aspects of the flora of northeastern Oaxaca*, and remains unpublished to this day. This field work gave Schultes his first experience in humid tropical vegetation, which reaches its northern limit on the eastern slopes of the mountains along the Gulf of Mexico. He travelled among six ethnic groups but concentrated on villages of the Chinatecs and Mazatecs. After documenting the uses for each plant with herbarium specimens (which are deposited in the Economic Herbarium of Oaks Ames (ECON), now part of the HUH), he made the appropriate taxonomic determination, searched each name's essential synonymy, cited the bibliographic references, and recorded its geographic range. All the ethnobotanical important plants are arranged according the Engler-Gilg classification system.

He emphasized plant uses; for each species he discussed it fully, compared it with reported uses in other parts of Mexico and tropical America, as well as correlated it with uses among pre-Hispanic peoples of Mexico. In the case of medicinal, narcotic, and poisonous plants, he listed the known phytochemical constituents and their bioactive properties. He acknowledged that ethnobotanical studies in tropical America was in an "infantile stage." He intended the Oaxacan study to be the first in a series of tropical American ethnobotanies that would parallel the Southwestern American ethnobotanies being developed by Edward Castetter (1944). Schultes' comparative studies demonstrated that 31% of plants in northeastern Oaxaca were used in the same way by Aztecs and Mayans, 54% of medicinal plants possessed known bioactive principles, and 73% of economic plants were employed for similar purposes. He explained these patterns in terms of the debate between diffusion versus independent experimentation. A typescript at the end of the thesis is an "ethnobotanical monograph" entitled *A contribution to our knowledge of Rivea corymbosa,* which was latter published by the Botanical Museum.

RES' legacy to ethnobotany as a whole took root in his Mexican studies. His structure and focus concentrated on history and narcotic plants. His mining of documents and specimens afforded unique perspective for his personal field work. History was a living part of his contemporary field activities. He added value to his predecessors' contributions, illuminating things that were hidden from his colleagues. For example, his enlightenment of the botanical locality of "chinantla," which was located in his field site (Schultes 1941), clarified the plant collections of German Karl Theodor Hartweg made a century earlier (McVaugh 1970). This passion for botanical history bore full fruit with his dedication later in life to the contributions of Richard Spruce in the Amazon.

His undergraduate work on peyote (*Lophophora williamsii*) introduced him to the products and processes of Mesoamerican ethnobotanical diffusion. His immersion into the geocultural region broadened his perspective on the diversity of this area of plant-human interaction. The unidentifiable fungal sample sent in 1938 by Blas Pablo Reko to Harvard University provided a key that unlocked the door of narcotic plants that dominated the rest of his life. This Austrian resident of Mexico, along with his cousin, Victor A. Reko, published in Europe their research on the "magic plants" of Mexico (Reko 1935, 1936) and introduced RES to Oaxaca. For Schultes, this opportunity not only provided the original research upon which to write his doctoral thesis, but also gave him the personal experience to confirm or rectify William E. Safford's interpretations of sacred Mexican plants (Safford 1915, 1917, 1922) (Table 1). For two of the five major plants (piceitl, tlapatl, peiotl, ololiuhqui, and teonanacatl), RES successfully defended his corrections of Safford's erroneous identifications. In the case of teonanacatl, RES obtained the first identifiable specimens

| Ethnotaxon | Safford | Schultes |
|---|---|---|
| Piceitl | *Nicotiana rustica* | *Nicotiana rustica* |
| Tlapatl | *Datura* spp. | *Datura* spp. |
| Ololiuhqui | *Datura* spp. | *Rivea corymbosa.* |
| Peiotl | *Lophophora williamsii* | *Lophophora williamsii* |
| Teonanacatl | *Lophophora williamsii* | *Psilocybe* spp. |

Table 1. Sacred plants of the Mexico: comparison of taxonomic determinations made by Safford and Schultes.

of three species of *Psilocybe*; subsequent investigations documented that teonanacatl consists of 24 species of Basidiomycetes (Schultes, 1978).

After defending his doctoral thesis in 1941, Schultes accompanied Paul C. Mangelsdorf to Mexico as part of the Rockefeller Foundation's Agricultural Program that lead to the inventory and improvement of maize in its center of origin and domestication. That same year, the National Research Council awarded Schultes a grant to study curare in South America. This opportunity permitted him to migrate southward, immerse himself in the tropical flora and their peoples, and escape Mexico that was becoming a 'crowded field.'

Between 1943 and 1957, few of Schultes' publications dealt with Mexico; afterwards Mexican data appeared imbedded in his general publications. Nonetheless, he encouraged others to follow his pioneering leads. R. Gordon Wasson began his ethnomycological field work in Oaxaca in 1953. His courteous association with RES is reflected in the donation of the "Tina and Gordon Wasson Ethnomycological Collection" to the Botanical Museum. RES also promoted Wasson's pioneering work on *Salvia divinorum*, actualization of the status of ololiuhqui, and novel interpretation of pre-Hispanic statue of xochilipilli and the role of 'flowers' in Nahuatl culture (Wasson 1962, 1963, 1973).

Schultes actively stimulated Mexican scientists to partake in the Harvard experience. As part of the Mexican maize program, Efraín Hernández Xolocotzi, considered to be the 'padre de la Etnobotánica en México' and a Distinguished Economic Botanist of the Society for Economic Botany, completed his master's program at Harvard (Hernández 1971). Antonio Marino Ambrosio received his doctoral degree there, with a thesis on the pulque agaves of Mexico (Marino 1966) and went on to become a high level government official in Mexico's Ministry of Agriculture and Hydrolic Resources (SARH). Jose Luis Díaz, a neurobiologist at the National Autonomous University of Mexico (UNAM), was a visiting research scientist at the Botanical Museum and worked on Mexican psychoactive plants (Díaz 1975).

After the 1940s, Schultes visited Mexico only twice as far as can be determined (to promote Spanish editions of his books and to visit the Seri Indians in Sonora with Richard Felger). He drew on his notes and memories to synthesize the geocultural phenomenon of hallucinogenic plants (Schultes 1977). These ideas were preceded by his compilation of data on the topic from around the world (Schultes and Hofmann 1973) and became available to Spanish-reading public later (Schultes and Hofmann 1982; Schultes 1982).

RES "baptized" his graduate students with experiences in South America. Although two of his students (Timothy Plowman and Tom E. Lockwood) conducted short field trips in Mexico after receiving their doctoral degrees, Robert Bye was the only student to have carried out field work in Mexico as part of his Harvard graduate program under the direction of Schultes. After being "reborn" in the Amazon, Bye returned to Harvard and continued the curatorial tasks assigned by RES that consisted of determining the taxonomic identifications and background of plant specimens transferred from the Peabody Museum of American Archaeology and Ethnology to the Botanical Museum. This Mexican collection was made by Edward Palmer during the 1870s in San Luis Potosí as part of his pioneering ethnobotanical work (Bye 1979) and it stimulated Bye to conduct field work in another of Palmer's study sites, the Sierra Tarahumara of Chihuahua. Bye continues his ethnobotanical and taxonomic investigations in various parts of Mexico to this day.

### CONCLUSION

Richard Schultes' academic progression from the campus in Cambridge, under the tutelage of Oakes Ames, to the Amazon traversed Mexico, where he left an unforgettable mark. There, he was able to meld his field experiences with historically puzzling records that baffled earlier scholars. His comparative approach to ethnobotany and economic botany, combined with a solid taxonomic foundation and interdisciplinary perspective, blossomed in Mexico and fructified in Colombia, the two major centers of "aboriginal" use of hallucinogens. His original contributions from Oaxaca served as the scientific foundation of Mexican ethnobotany and solidified his approach to the study of plant-people interactions of the Americas. He encouraged Mexican and foreign scientists to pursue ethnobotanical investigation in Mexico, although he never returned to the field in Mexico after 1941. As a consequence, RES's experiences solidified his academic character and Mexican ethnobotanical research developed with a firm footing.

### ACKNOWLEDGMENTS

I thank the many people have generously assisted my exploration of the role Schultes played in Mexican ethnobotany. At the Instituto de Biología, UNAM, I acknowledge in particular Edelmira Linares, Georgina Leite, and Armando Butanda. Members of the Harvard University Herbaria Libraries and Archives staff were always attentive to my strange requests: Judith Warnement, Lisa DeCesare, Gretchen Wade, and Peg Richards. The staff of the Harvard University Herbaria greatly facilitated my study of the files and specimens: Don Pfister, Robert Cook, James Macklin, Emily Wood, Stephanie Zabel, Henry Kesner, Jessica Dolan, and Melinda Peters. The award of the Antonio Madero Fellowship by the David Rockefeller Center for Latin American Studies, Harvard University, gave me an opportunity to study a sampling of Schultes' documents.

### LITERATURE CITED

Bye, R. 1979. An 1878 ethnobotanical collection from San Luis Potosi: Dr. Edward Palmer's first major Mexican collection. *Econ Bot* 33:135–162.

Castetter, E.F. 1944. The domain of ethnobiology. *Am Nat* 78:158–170.

Díaz, J.L. (ed.). 1975. *Etnofarmacología de Plantas Alucinógenicas Latinoamericanas*. Cuadernos Científicos CEMEF 4, México, DF.

Dupree, A.H. 1959. *Asa Gray*. Harvard University Press, Cambridge, Mass.

Hernández Xolocotzi, E. 1971. *Apuntes sobre la Exploracion Etnobotanica y su Metodología*. Chapingo, Estado de México: Colegio de Postgraduados.

Herrera, T., M.M. Ortega, J.L. Godínez, and A. Butanda. 1998. *Breve Historia de la Botánica en México*. Fondo de Cultura Económica, México, DF.

Marino Ambrosio, A. 1966. *The Pulque Agaves of Mexico*. Cambridge, MA: Doctor of Philosophy Thesis, Harvard University.

McVaugh, Rogers. 1970. Introduction to the facsimile reprint of George Bentham's *Plantae Hartwegianae*, with two maps. J. Cramer Verlag, Lehre, Germany.

Prance, G.T. 2001. Richard Evans Schultes (12 January 1915–10 April 2011): a tribute. *Econ Bot* 55:347–362.

Reko, V.A. 1935. Nanácatl: The "Delirum Fungus." *Pharm Monastsh* 16:29–31.

Reko, V.A. 1936. *Magische Gifte*. Stuttgart: Ferninand Enke Verlag.

Safford, W.E. 1915. An Aztec narcotic. *J Hered* 6:291–311.

Safford, W.E. 1917. Narcotic plants and stimulants of the ancient Americans. *Annual Report, Smithsonian Institution for 1916*. Pp. 387–424.

Safford, W.E. 1922. Daturas of the Old World and New: an account of their narcotic properties and their use in oracular and initiatory ceremonies. *Annual Report, Smithsonian Institution for 1920*. Pp. 537–567.

Schultes, R.E. 1941. The meaning and usage of the Mexican place-name "Chinantla." *Botanical Museum Leaflets* 9:101-116.

Schutles, R.E. 1977. Mexico and Colombia: Two major centers of aboriginal use of hallucinogens. *J Psychedel Drug* 9:173–176.

Schultes, R.E. 1978. Evolution of the identification of the sacred hallucinogenic mushrooms of Mexico, Pp. 26–43 in J. Ott and J. Bigwood (eds.), *Teonanacatl: Hallucinogenic Mushrooms of North America*. Madrona Publishers, Inc., Seattle, Wash.

Schutles, R.E. 1982. *Plantas Alucinógenicas*. México, DF: Prensa Médica Mexicana.

Schutles, R.E., and A. Hofmann. 1973. *The Botany and Chemistry of Hallucinogens*. Charles C. Thomas Publishers, Springfield, Ill.

Schultes, R.E., and A. Hofmann. 1982. *Plantas de los Dioses*. México, DF: Fondo de Cultura Económica.

Wasson, R.G. 1962. A new Mexican psycotropic drug from the mint family. *Botanical Museum Leaflets* 20:77–84.

Wasson, R.G. 1963. Notes on the present status of ololiuhqui and the other hallucinogens of Mexico. *Botanical Museum Leaflets* 20:161–212.

Wasson, R.G. 1973. The role of 'flowers' in Nahuatl culture: a suggested interpretation. *Botanical Museum Leaflets* 23:305–324.

# A Standard Protocol for Gathering Palm Ethnobotanical Data and Socioeconomic Variables across the Tropics

Rodrigo Cámara-Leret,[1] Narel Y. Paniagua Zambrana,[1,2] and Manuel J. Macía[1]

This paper describes a protocol for collecting information on palm ethnobotany and related socioeconomic variables in rural communities across the tropics. The steps to follow when conducting quantitative ethnobotanical fieldwork are presented chronologically: 1) selection of study communities, 2) preparation of materials and permits, 3) planning work at the community, 4) community census, 5) selection of informants, 6) types of interviews for gathering ethnobotanical data and socioeconomic variables, and 7) returning information to the communities. Although this protocol was developed and tested in northwestern South America, it can be used for comparing palm use patterns in any country, ecoregion, habitat, human group and use categories across the tropics.

## Introduction

In the last decades, ethnobotanists have broadened the discipline's methods and goals resulting in a shift from purely descriptive to more quantitative approaches (Carneiro 1978, Trotter and Logan 1986, Prance et al. 1987, Johns et al. 1990, Phillips and Gentry 1993, Phillips et al. 1994, Galeano 2000, Macía et al. 2001, Collins et al. 2006, Reyes-García et al. 2007, Vandebroek 2010). Within the framework of the PALMS project (http://www.fp7-palms.org), we designed and tested (>2000 interviews) during almost two years of fieldwork in northwestern South America (Colombia, Ecuador, Peru, and Bolivia) the protocol we present in this paper for gathering palm ethnobotanical data and related socioeconomic variables. Although many articles and books discussing methodology exist (e.g., Martin 1995, Alexiades 1996, Höft et al. 1999, Reyes-García et al. 2007), we chose to disseminate this protocol for its precise focus on a specific group of plants: Arecaceae. To our knowledge, this is the first regional initiative in quantitative palm ethnobotany (covering several countries, ecoregions and human groups). Hence, we propose to extrapolate this same methodology to other areas to enable global comparisons of palms use patterns.

Palms are an ideal ethnobotanical group of plants to be studied because they are easily recognized by people worldwide, there are many useful species, and they are very ecologically important (Henderson et al. 1995, Macía 2004, De la Torre et al. 2009, Balslev et al. 2011). Quantitative ethnobotanical studies on palms are scarce and have mostly been focused on topics such as the relationships between palm use and palm diversity and abundance (Byg et al. 2006, De la Torre et al. 2009), socioeconomic factors (Byg and Balslev 2001a, b; Paniagua et al. 2007), palm ecological and morphological characteristics (Byg et al. 2006, De la Torre et al. 2009), palm cultivation and management (Byg and Balslev 2006) and use categories and subcategories, ecoregions, countries and human groups in northwestern South America (Macía et al.

---

[1] Departamento de Biología, Área de Botánica, Edificio de Ciencias Biológicas, Universidad Autónoma de Madrid, Calle Darwin 2, E-28049, Madrid, Spain. Email: rcamaraleret@gmail.com.

[2] Herbario Nacional de Bolivia, Universidad Mayor de San Andrés, Casilla 10077 – Correo Central, La Paz, Bolivia. Email: nyaroslava@yahoo.es.

2011). The compilation of information using database management systems has produced important reference works (Balick and Beck 1990) and quantitative reviews (Macía et al. 2011), providing a solid base for highlighting research needs in the future.

Palm ethnobotanical fieldwork has been facilitated by the existence of field guides (Henderson et al. 1995, Dransfield et al. 2006, Galeano and Bernal 2010), and during interviews the use of photographs has been recommended (Nguyen 2003, Thomas et al. 2007), although limitations for some taxa exist. Since the first field guide to tropical palms was published (Wallace 1853, Knapp et al. 2002), there have been many comprehensive advances in palm taxonomy. In the Neotropics, palm taxonomy is sufficiently well understood (Henderson 1995, Henderson et al. 1995, Pintaud et al. 2008, Lorenzi 2010, Galeano and Bernal 2010, Henderson 2011) although new species are still described (Henderson 2005, 2011; Lorenzi et al. 2010). For some parts of Africa, the palm flora is well worked out (Dransfield 1986, Tuley 1995, Sunderland 2007). The palms of Madagascar have been recently monographed (Dransfield et al. 2006), but still new palm species are being discovered (Rakotoanirivo et al. 2007, Dransfield et al. 2008b, Rakotoarinivo 2008). Taxonomic studies in Asia lag behind, which is likely due to high species diversity. Southern Asia is apparently better understood, with a guide of its species recently published (Henderson 2009). The Malesian palm flora, the richest on earth with 992 estimated species in 50 genera (Dransfield et al. 2008), lacks a modern taxonomic treatment. This is often also the case at the country level. Indonesia, for instance, has 477 known palm species (Johnson 1991), but a detailed flora exists only for the rattans of Borneo (Dransfield and Patel 2005). Likewise, at the island level only a generic field guide of New Guinea palms has been published (Baker and Dransfield 2006). Considering Asia's unrivaled species diversity and its rich ethnic diversity in comparison to the Neotropics (the island of New Guinea alone harbors over 1000 different ethnic groups), palm ethnobotanical studies are needed in most countries and regions.

As suggested by numerous ethnobotanists, comparative studies and their replication in different parts of the world are warranted to identify similarities and differences in the use of plant resources and for the development of a well-formulated theory for ethnobotany (Albuquerque and Hanakazi 2009). In northwestern South America, palm ethnobotanical research priorities have been highlighted by Macía et al. (2011). In Asia, comparative palm ethnobotany is practically nonexistent (but see Johnson 1991, 2011), and no comprehensive bibliographical review exists as in the Neotropics of palms use besides that of rattans (Dransfield and Manokaran 1993, Wulijarni-Soitjipto and Danimihardja 1995).

For the reasons listed above, we present a protocol with the goal of contributing to a more profound and unified ethnobotanical research approach of palm use patterns throughout the tropics. As an example, based on our work in northwestern South America (Paniagua et al. 2010), among the aims and research questions that can be addressed are:

1) Comparatively describe the diversity of useful palms and their use patterns in the tropical humid forests at different scales:
- Countries: e.g., Colombia, Ecuador, Peru, and Bolivia

- Ecoregions: e.g., Amazonia, Andes, and the Pacific humid coast (Chocó)
- Habitats: e.g., Amazon terra firme and inundated forests, humid montane forest, and Chocó forests
- Human groups: e.g., indigenous, afroamerican, mestizo, and settler (colonos)
- Use categories and subcategories

2) Analyze the factors that determine distribution and transmission of knowledge in the use of palms from an ecologic, socioeconomic, cultural, and historical perspective.
3) Study the relationship between palm use patterns and their morphologic, ecologic, and phylogenetic characteristics.
4) Determine if the characteristics of palm use are related to their diversity and abundance in the forests near human settlements, using quantitative inventories.
5) Assess the influence of accessibility to urban centers and markets on the use of palms.
6) Evaluate the degree of management, cultivation, and commercialization of palms and palm products.

## Protocol

The steps to follow when conducting quantitative ethnobotanical fieldwork are presented chronologically and include: 1) selection of study communities, 2) preparation of materials and permits, 3) planning work in the communitites, 4) community census, 5) selection of informants, 6) types of interviews and ethnobotanical data gathering, and 7) returning information to the communities.

### 1. Selection of communities

Considering the compilation of palm ethnobotanical knowledge across different ecoregions, forest types and several ethnic groups in varying degrees of accessibility, we propose a protocol and sample size taking into account all these criteria at local and broad scales:

#### *A - Ecoregions and Forest types*

Palm communities vary greatly from one ecoregion to another (Bjorholm et al. 2005) and from one forest type to another (Kristiansen et al. 2011). In northwestern South America, to understand how ethnobotanical knowledge and palm use are related to ecological traits (palm floristic composition, individual abundances, species morphology, and phylogeny) and to socioeconomic factors (informant age, gender, status, ethnicity), communities found in the proximity of three target ecoregions were selected: 1) Amazonian terra firme, floodplain and swamp forests, 2) humid Andean montane forests and 3) Chocó forests. In each ecoregion, at least one locality was selected where two different ethnic groups share the same palm flora.

To compare ethnobotanical information with ecological parameters, palm transect data were gathered in all target areas. A detailed protocol on how to conduct palm transects can be found in Balslev et al. (2010).

#### *B - Ethnic Origin*

Intercultural or intercommunity ethnobotanical comparisons are desirable but still scarce (Benz et al. 2000), and those focusing on palms are practically absent (e.g., De la Torre et al. 2009). For instance, in northwestern South America, only 49% of Amazonian, and 29% of Andean, indigenous groups count with palm ethnobotanical information, and even this information is only preliminary for many ethnicities (Macía et al. 2011). To compare palm use between different human groups, in each study locality the aim was to work with communities where one human group is dominant. In our example of northwestern South America, the following human groups were studied: 1) indigenous, 2) afroamerican, 3) mestizo, 4) settler (colono) and 5) mixed communities (where several of the human groups described coexist).

*C - Accessibility*

Previous studies have shown that less access to markets and large cities, linked to poorer market integration, insufficient government services, and infrastructure (schools, health, electricity, telephone) associates with a greater preservation of local ethnobotanical knowledge (Benz et al. 2000, Byg and Balslev 2001a, 2004; Byg et al. 2007, Crepaldi and Peixoto 2010, De la Torre et al. 2012). To compare the correlation between accessibility and ethnobotanical knowledge throughout the four countries in northwestern South America to test previous conclusions, we selected one to several communities, where each ethnic group was dominant on the basis of differential proximity to cities or markets. Accessibility was considered as the combination of 1) distance from the communities to larger cities and towns (in kilometers) and 2) transportation type (see Appendix 1).

*D - Sample size*

To compare local ecological knowledge between gender and age groups as suggested by previous researchers (Begossi 1996, Caniago and Siebert 1998, Begossi et al. 2002, Reyes García et al. 2007), we interviewed (in most localities if possible) at least 87 informants in five age categories per locality (see Table 1 for gender and age group distribution). To reach the number of informants planned, on occasions we worked in only one community but in other cases we had to visit as many communities as necessary (all of the same ethnic group) to meet this figure. In our research, a large sampling size was chosen since effective sampling size is a prerequisite for determining the factors that explain the use and knowledge of plants regionally and for a most significant statisticaltesting (Albuquerque and Hanakazi 2009).

2. Preparation of work materials and permits before travelling

Prior to conducting ethnobotanical research, the first step is to obtain all research permits and pertinent authorizations from official institutions and from the communities where the study is to take place, and to establish prior informed consent.

To facilitate fieldwork the following material needs to be prepared before travelling:

| GENDER/AGE CATEGORY | 18–30 years | 31–40 years | 41–50 years | 51–60 years | >60 years |
|---|---|---|---|---|---|
| Male | 8 | 9 | 9 | 9 | 9 |
| Female | 8 | 9 | 9 | 9 | 9 |
| TOTAL | 16 | 18 | 18 | 18 | 18 |

Table 1. Distribution per locality of the number of informants in relation to gender and age categories.

- Copies of the project description, research and collecting permits and correspondence with permitting entities and community organizations / community leaders
- List of the names and contact information of the community leaders and other important people from the selected study area
- List of palm species to be found in the area with their vernacular names in different languages
- Printed questionnaires as guidelines for ethnobotanical and socioeconomic data gathering (Appendices 1–7)
- A field guide with color photographs of the palms is highly recommended. See Appendix 8 for a list of the most important tropical palm field guides.

3. Planning work in the communities

*A - Arrival at the community*

Once in the community, the researchers will contact first the local leader(s) to explain the scope of the research and the activities planned during the researchers' visit, in order to set a time for organizing a meeting with all the community members, as prerequisite for obtaining prior informed consent. This meeting is preferably conducted at the time of day that is most appropriate to ensure that all community members are present.

The establishment of confidence between researchers and community members is a process that requires time. Prior to beginning ethnobotanical interviews, it is important to acknowledge and familiarize oneself with the daily activities and with the people who live in the community. For that reason, this period is dedicated to complementary activities such as gathering general information on the socioeconomic and historical background of the community with the local leader(s), and additional meetings (if they are requested) after obtaining prior informed consent in which the research work is planned. Developing these activities properly will allow for greater possibilities in the next work phase and most important one, the ethnobotanical interviews and collection of palm specimens.

*B - Interview with the local leader(s)*

This interview can be done (at least in part) on the day of arrival to the community. In the case of time constraints, it is important to program this activity as a priority, as it must be done before any other tasks. Activities during the interview include: 1) Gathering socioeconomic and historical information of the community (questionnaire in Appendix 1); 2) Gaining insight on the community member's routines in order to plan interview times more effectively; 3) Elaboration of a synthetic community map, with approximate location of the houses (to be used as support of the census, for a example see Appendices 1 and 2); 4) Drawing a simple vegetation map showing the different forest types (to plan the field trips with the expert informants); 5) Names of members with higher knowledge on palm uses, (to be considered and suggested as possible expert informants during the community meetings).

*C - Meeting and presentation of research to the community*

In the meeting, the researchers need to again explain the objectives, methodology and outputs of the research to all community members in a simple fashion. Additionally, details of all activities to be developed within the community must be provided, including 1) interviews with the leaders, 2) community census, 3) selection of expert and general informants, 4) fieldwork with expert informants, 5) collection of botanical material, 6) interviews within the community and schedule of visits in the different houses, 7) economic matters, 8) outputs of the research to the community and to collaborators.

After the meeting, prior informed consent should be established and signed on the equitable benefit sharing arising from research, along the lines of the Convention of Biological Diversity articles 8j and 15 (http://www.cbd.int) and its future protocol about access to genetic resources and benefit sharing. In addition, national legislation on access to genetic material and traditional knowledge and benefit sharing must be fulfilled. If the community does not wish to participate in the research, this must be respected.

4. Community census

It is recommended to elaborate the community census and community map accompanied by a community leader or other member(s) (including children) who can facilitate the conversations and help obtain the correct names of the members inhabiting each house, even if these are away. This activity is a first chance to acquaint oneself with the local inhabitants, for both parts to initiate friendships (and gain more confidence) and for the researcher to gain first hand information needed to select the final interviewees. All houses are to be visited and information gathered using the questionnaire provided in Appendix 2. In the case of working in towns or large communities with numerous inhabitants and houses, the community census may be too complex and time consuming to complete and therefore can be obviated.

5. Selection of informants

To comprehensively document palm ethnobotanical lore in the communities our interview strategy is based on two types of informants: 1) Seven expert informants, selected mainly by members of the community for their greater knowledge on the forest around the community, the palm species, their uses, processing, harvesting and management; and 2) Eighty general informants, chosen by the researchers to obtain information on how palm use knowledge varies within members in the community. This type of approach using various kinds of informants has been recommended previously (Davis and Wagner 2003, Vandebroek 2010, Ruelle and Kassam 2011).

*A - Expert informants*

Recognizing the necessity of employing a systematical methodological approach when identifying expert informants (Davis and Wagner 2003), these are selected based on information provided by the community leader(s), community meetings and talks with members of the community (Crepaldi and Peixoto 2010). Expert informants are usually over 35 years age, because they are more experienced, and the participation of women is encouraged.

Working with the expert informant(s) consists on three fundamental tasks: 1) walking in different forest types neighboring the communities in search of all palm species present, 2) collecting palm specimens, and 3) interviewing the expert informant(s) (see questionnaire in Appendix 4). The time suggested to work with each expert informant is half to a full day depending on the palm diversity of the locality and the expert informant's ethnobotanical knowledge. Seven experts accompany the researchers in different days to document as many vernacular names as possible during fieldwork.The use of vernacular names is the basis for interviewing the majority of informants, and effort must be made during the walk in the woods with the expert(s) to assure matching them with actual palm species. Vernacular names can be registered using a voice recorder and transcribed later with the aid of bilingual teachers back in the community. It is recommended to dedicate sufficient time to collect duplicates of botanical material so as to distribute material to national and international herbaria and palm taxonomic experts. Details on preparing palm specimens can be found in Dransfield (1986) and Balslev et al. (2010). During the walk in the woods, all data from the questionnaire (Appendix 4) should be filled for every species found but also for those the informant knows from other areas.

*B - General informants*

For their selection, the information gathered during the population census in the community prior interviewing is used. It is also highly recommendable to work with a local guide that facilitates finding and interviewing informants within the community. Depending on the size of the community, a person from each family is selected, trying to interview at least 80% of the families that live in one community. If the minimum number of interviews required is not fulfilled in one community, the distribution of informants per gender and age range would be among several communities, depending on population.

To evaluate the gender differences in palm use knowledge a proportion of 50% female and 50% male informants in a given locality (or community) was targeted. However, it is important to note that in certain localities these proportions are difficult to obtain (e.g., the expert informants are often male, insufficient number of women above age 61 in a locality, etc.). The relationship between the informant's age and the knowledge on palm use is studied to analyze the transmission of palm ethnobotanical knowledge between generations, grouping the latter into five categories (Table 1). Efforts are made to achieve an even grouping of informants according to the age ranges.

6. Types of interviews and ethnobotanical data gathering

The following general aspects should be considered during the interviews:

- The interviewee must be informed that the interview will be written down or recorded, and that the information gathered is private and just for scientific use.
- To compare individual knowledge among informants, we suggest working in a calm setting where the interviews must be realized with one person at a time, avoiding the presence of others around. Yet, this ideal case is not always achieved and we recognize the difficulty of working with only one informant, especially with ethnic groups where women are culturally banned from speaking to outsiders. Thus sometimes, it might be more realistic to interview couples, and through comparative questions assess the differences in knowledge between informants.

- The interviews should be done in a place where the informant feels comfortable, which is usually their home.
- It is important not to rush answers during the interviews, even if time is limited, but also to lead the interview and avoid different topics other than palm ethnobotany.
- Sensibility and respect are always important, even more when informants prove shy or uncomfortable with certain questions or they do not have answers for the questions.
- In the case interviewers require translators, these must be competent and well versed in both languages.

To gather ethnobotanical information and socioeconomic variables from the informants, two types of interviews are utilized (Martin 1995, Alexiades 1996, Cunningham 2001):

*A) Structured interviews*

Structured interviews are based on a fixed number of standardized questions used for all informants. This methodology allows for greater control of the answers received. These interviews are conducted with: 1) the community leader to obtain socioeconomic and historical information of the community (Appendix 1) and with all informants (expert and general) for gathering personal socioeconomic information (Appendix 3).

*B) Semi-structured interviews*

Semi-structured interviews are composed of a list of open-ended questions, to obtain greater responses than in the previous case. These interviews have a greater degree of flexibility, as from certain responses new questions can arise. At the end of the interview all proposed questions must be answered. These interviews are to be used with expert and general informants.

To obtain information from expert and general informants on the uses and products of all plant species targeted, uses according to each plant part are grouped in 10 ethnobotanical categories and various subcategories (Appendix 5). A palm use is defined as in Macía et al. (2011): the use associated to a use category and a use subcategory for a specific plant part (Appendix 6).

These types of interviews are developed in various phases:

1. Field work:

With each of the expert informants all palm species in the different vegetation types surrounding the communities are visited during the walks in the forest, and it is highly recommended to collect the different species found (see section 5A.).

2. Work in the expert informants' houses

A second directed interview is conducted asking about the existence and utilization of palm species that grow in common in all target areas, in our test case in the humid tropical forests of Colombia, Ecuador, Peru, and Bolivia (Appendix 7); this list of species is based on a bibliographic revision of palm uses and distribution (Macía et al. 2011). Many of the common species will already have been found during fieldwork with the expert informants, and questions should focus only on the remaining species. Two levels of priority are proposed for the registration of the information, based on the probability of finding species in the field: 1) High priority: abundant and conspicuous species, easy to recognize in the forests and which have a vast range of uses or a certain outstanding use; 2) Medium priority: species which can be locally abundant, but are less frequent in the forests and have a smaller range of uses. By defining levels of priority it is possible to evaluate the relationships between local availability and local perception of the importance of the resource, contrasting at broader scales results of previous ethnobotanical studies (Phillips and Gentry 1993, Galeano 2000, Macía et al. 2001, Byg et al. 2006, De la Torre et al. 2009).

3. Revisionary work

Once field interviews are concluded with all expert informants, some of them may be interviewed again in the community to obtain ethnobotanical information on those palm species not encountered with them during fieldwork. The overall objective is that all expert informants contribute with use information on all palm species existing in the region.

4. Work in the general informants' houses

When all experts have been interviewed, general informants are visited in their respective houses and interviewed using as reference the list of useful palms gathered with the experts. This list may be complemented showing the interviewees photographs (Foster et al. 2004, Galeano and Bernal 2010) or fresh palm parts (leaves, fruits, seeds) if available. General informants will be asked about all palm species growing near the community, although the questionnaire is simplified with respect to that of the experts. If any of the general informants shows profound knowledge he/she may be considered as an expert informant and interviewed with greater detail using the expert questionnaire. Additionally, at the end of the questionnaire informants are asked about the existence and utilization of palm species of high priority (species growing in common in the forests of Colombia, Ecuador, Peru, and Bolivia)and of medium priority (only if they are present in the forests near the community)(Appendix 7), if these have not been mentioned.

*C) Informal interviews*

All casual conversations that may arise besides interviews (during community meetings, elaboration of the census, etc.) are grouped under this category. It is recommended that information from these conversations be registered in field books, mentioning the informants' name.

7. Return of information to the communities

Once work is completed in the communities, the registered ethnobotanical information should be returned in a manner appropriate for the community. The way of returning information, initially agreed upon during the community meetings, may change as community members and informants get more familiar with the study approach during fieldwork. The final form and scope of returning the research results to the community should be discussed and defined in a final community meeting.

Acknowledgments

The authors thank H. Balslev and R. Bussmann for valuable comments for improving the manuscript. Our gratitude goes to W. Baker, J. Dransfield, M. Balick and K. Herrera Hurtado for their contributions to improving the list of field guides. This ongoing research is part of the Palm Harvest Impacts in Tropical Forests PALMS project funded by the European Union as part of the 7th Framework Program (Grant No. 212631).

## LITERATURE CITED

Alexiades, M.N. 1996. Collecting Ethnobotanical Data: An Introduction to Basic Concepts and Techniques. Pp. 53–94 in Alexiades, M.N. (ed.), *Selected Guidelines for Ethnobotanical Research: A Field Manual.* The New York Botanical Garden, New York.

Albuquerque, U.P. de, and N. Hanazaki. 2009. Five Problems in Current Ethnobotanical Research-and Some Suggestions for Strengthening Them. *Hum Ecol* 27:653–661.

Baker, W.J., and J. Dransfield. 2006. *Field Guide to the Palms of New Guinea.* Kew Publishing, Royal Botanic Gardens, Kew, Richmond.

Balick, M.J., and H.T. Beck (eds.). 1990. *Useful Palms of the World: A Synoptic Bibliography.* Columbia University Press, New York.

Balslev, H., H. Navarrete, N. Paniagua-Zambrana, D. Pedersen, W. Eiserhardt, and T. Kristansen. 2010. El uso de transectos para el estudio de comunidades de palmas. *Ecología en Bolivia* 45(3):8–22.

Balslev, H., F. Kahn, B. Millan, J.C. Svenning, T. Kristiansen, F. Borchsenius, D. Pedersen, and W. Eiserhardt, 2011. Species Diversity and Growth Forms in Tropical American Palm Communities. *Bot Rev* 77:381–425.

Begossi, A. 1996. Use of Ecological Methods in Ethnobotany: Diversity Indices. *Econ Bot* 50:280–289.

Begossi, A., N. Hanazaki, and J.Y. Tamashiro. 2002. Medicinal Plants in the Atlantic Forest (Brazil): Knowledge, Use, and Conservation. *Hum Ecol* 30:281–299.

Benz, B.F., E.J. Cevallos, M.F. Santana, A.J. Rosales, and M.S. Graf. 2000. Losing Knowledge about Plant Use in the Sierra de Manantlan Biosphere Reserve, Mexico. *Econ Bot* 54:183–191.

Bernal, R., C. Torres, N. García, C. Isaza, J. Navarro, M.I. Vallejo, G. Galeano, and H. Balslev. 2011. Palm Management in South America. *Bot Rev* 77(4):607–646.

Bjorholm, S., J-C. Svenning, F. Skov, and H. Balslev. 2005. Environmental and Spatial Controls of Palm (Arecaceae) Species Richness across the Americas. *Global Ecol Biogeogr* 14:423–429.

Byg, A., and H. Balslev. 2001a. Diversity and Use of Palms in Zahamena, Eastern Madagascar. *Biodivers Conserv* 10:951–970.

———. 2001b. Traditional Knowledge of *Dypsis fribrosa* (Arecaceae) in Eastern Madagascar. *Econ Bot* 55(2):263–275.

———. 2004. Factors Affecting Local Knowledge of Palms in Nangaritza Valley in South-Eastern Ecuador. *J Ethnobiology* 24:255–278.

———. 2006. Palms in Indigenous and Settler Communities in Southeastern Ecuador: Farmer's Perceptions and Cultivation Practices. *Agroforest Syst* 67:147–158.

Byg, A., J. Vormisto, and H. Balslev. 2006. Using the Useful: Characteristics of Used Palms in South-Eastern Ecuador. *Environment, Development and Sustainability* 8:495–506.

———. 2007. Influence of Diversity and Road Access on Palm Extraction at Landscape Scale in SE Ecuador. *Biodivers Conserv* 16:631–642.

Carneiro, R.L. 1978. The Knowledge and Use of Rain Forest Trees by the Kuikuru Indians of Central Brazil. Pp. 201–216 in R.I. Ford (ed.), *The Nature and Status of Ethnobotany.* University of Michigan Press, Ann Arbor, Michigan.

Collins, S., X. Martins, A. Mitchell, A. Teshome, and T.A. Johns. 2006. Quantitative Ethnobotany of Two East Timorese Cultures. *Econ Bot* 60(4):347–361.

Cook, F.E.M. 1995. *Economic Botany Data Collection Standard.* Royal Botanic Gardens, Kew, London.

Crepaldi, M.O.S., and A.L. Peixoto. 2010. Use and Knowledge of Plants by "Quilombolas" as Subsidies for Conservation Efforts in an Area of Atlantic Forest in Espírito Santo State, Brazil. *Biodivers Conserv* 19:37–60.

Cunningham, A.B. 2001. *Applied Ethnobotany. People, Wild Plant Use and Conservation.* WWF, UNESCO,

Royal Botanical Gardens, Kew. Earthscan Publications, London and Sterling.

Davis, A., and J.R. Wagner. 2003. Who Knows? On the Importance of Identifying "Experts" When Researching Local Ecological Knowledge. *Hum Ecol* 31(3):463–489.

De la Torre, L., L.M. Calvo-Irabién, C. Salazar, H. Balslev, and F. Borchsenius. 2009. Contrasting Palm Species and Use Diversity in the Yucatan Peninsula and the Ecuadorian Amazon. *Biodivers Conserv* 18:2837–2853.

De la Torre, L., C. Cerón, H. Balslev, and F. Borchsenius. 2012. A Biodiversity Informatics Approach to Ethnobotany: Meta-Analysis of Plant Use Patterns in Ecuador. *Ecol Soc* 17(1):15.

Dransfield, J. 1986. A Guide to Collecting Palms. *Ann MO Bot Gard* 73(1):166–176.

Dransfield, J., and M. Patel. 2005. *Rattans of Borneo: An Interactive Key.* CD Rom. Kew Publishing, Royal Botanic Gardens, Kew, Richmond.

Dransfield J., and N. Manokaran (eds.). 1993. *Plant Resources of South-East Asia No. 6: Rattans.* Pudoc Scientific Publishers, Wageningen.

Dransfield, J., N.W. Uhl, C.B. Asmussen, W.J. Baker, M.M. Harley, and C.E. Lewis. 2008. *Genera Palmarum. The Evolution and Classification of Palms.* Kew Publishing, Royal Botanic Gardens, Kew, Richmond.

Dransfield, J., B. Leroy, X. Metz, and M. Rakotoarinivo. 2008. *Tahina* – A New Palm Genus from Madagascar. *Palms* 52(1):31–39.

Dransfield, J., H. Beentje, A. Britt, T. Ranarivelo, and J. Razafitsalama. 2006. *Field Guide to the Palms of Madagascar.* Kew Publishing, Royal Botanic Gardens, Kew, Richmond.

Foster R., H. Betz, M. Metz, T. Wachter, M. Giblin. 2004. *Palmas Communes del Peru.* Rapid Color Guide # 56 version 1.2. Environmental and Conservation Programs, The Field Museum, Chicago.

Galeano, G. 2000. Forest Use at the Pacific Coast of Chocó, Colombia: A Quantitative Approach. *Econ Bot* 54(3):358–376.

Galeano, G., and R. Bernal. 2010. *Palmas de Colombia. Guía de Campo.* Universidad Nacional de Colombia. Instituto de Ciencias Naturales, Bogotá.

Henderson, A. 1995. *The Palms of the Amazon.* Oxford University Press, New York.

———. 2009. *Palms of Southern Asia.* Princeton University Press, Princeton.

———. 2005. A Multivariate Study of *Calyptrogyne* (Palmae). *Syst Bot* 30:60–83.

———. 2011. A revision of *Geonoma* (Arecaceae). *Phytotaxa* 17:1–271.

Henderson, A., G. Galeano, and R. Bernal. 1995. *Field Guide to the Palms of the Americas.* Princeton University Press, Princeton.

Höft, M., S.K. Barik, and A.M. Lykke. 1999. *Quantitative Ethnobotany. Applications of Multivariate and Statistical Analyses in Ethnobotany.* People and Plants Working Paper 6. United Nations Educational, Scientific and Cultural Organization (UNESCO), Paris.

Johns, T., J.O. Kokwaro, and E.K. Kimanani. 1990. Herbal Remedies of the Luo of Siaya District, Kenya: Establishing Quantitative Criteria for Consensus. *Econ Bot* 3:1–10.

Johnson, D. (ed.). 1991. *Palms for Human Needs in Asia.* A.A. Balkema, Rotterdam.

———. 2011. *Non-Wood Forest Products 10/Rev. 1. Tropical Palms.* 2010 revision. FAO, Rome.

Knapp, S., L. Sanders, and W. Baker. 2002. Alfred Russel Wallace and the Palms of the Amazon. *Palms* 46(3):109–119.

Kristiansen, T., J-C. Svenning, D. Pedersen, W.L. Eiserhardt, C. Grandez, and H. Balslev. 2011. Local and Regional Palm (Arecaceae) Species Richness Patterns and Their Cross-Scale Determinants in the Western Amazon. *J Ecol* 99:1001–1015.

Lorenzi, H., L. Noblick, F. Kahn, and E. Ferreira. 2010. *Flora Brasileira (Arecaceae).* Editora Plantarum, Nova Odessa.

Macía, M.J. 2004. Multiplicity in Palm Uses by the Huaorani of Amazonian Ecuador. *Bot J Linn Soc* 144:149–159.

Macía, M.J., H. Romero-Saltos, and R. Valencia. 2001. Patrones de uso en un Bosque Primario de la Amazonía Ecuatoriana: Comparación Entre Dos Comunidades Huaorani. Pp. 225–249 in J.F. Duivenvoorden, H. Balslev, J. Cavelier, C. Grandez, H. Tuomisto, and R. Valencia (eds.), *Evaluación de Recursos Vegetales no Maderables en la Amazonía Noroccidental.* Institute for Biodiversity and Ecosystem Dynamics, Universiteit van Amsterdam, Amsterdam.

Macía, M.J., P.J. Armesilla, R. Cámara-Leret, N. Paniagua-Zambrana, S. Villalba, H. Balslev, and M. Pardo-de-Santayana. 2011. Palm Uses in Northwestern South America: A Quantitative Review. *Bot Rev* 77(4):462–570.

Martin, G.J. 1995. *Ethnobotany. A "People and Plants" Conservation Manual.* World Wide Fund for Nature. Chapman & Hall, London.

Nguyen, M.L. 2003. Comparison of Food Plant Knowledge between Urban Vietnamese Living in Vietnam and Hawai'i. *Econ Bot* 57(4):472–480.

Paniagua, N.Y., A. Byg, J.C. Svenning, M. Moraes, C. Grández, and H. Balslev. 2007. Diversity of Palm Uses in the Western Amazon. *Biodivers Conserv* 16:2771–2787.

Paniagua-Zambrana, N., M.J. Macía, and R. Cámara-Leret. 2010. Toma de Datos Etnobotánicos de Palmeras y Variables Socioeconómicas en Comunidades Rurales. *Ecología en Bolivia* 45(3):44–68.

Phillips, O., and A. Gentry. 1993. The Useful Plants of Tambopata Peru: II Additional Hypothesis Testing in Quantitative Ethnobotany. *Econ Bot* 47:33–43.

Phillips, O., A.H. Gentry, C. Reynel, P. Wilkin, and C. Gáldez-Durand B. 1994. Quantitative Ethnobotany and Amazonian Conservation. *Conserv Biol* 8(1):225–248.

Pintaud, J., G. Galeano, H. Balslev, R. Bernal, F. Borchsenius, E. Ferreira, J. de Granville, K. Mejía, B. Millán, M. Moraes, L. Noblick, F.W. Stauffer, and F. Kahn. 2008. Las Palmeras de América del Sur: Diversidad, Distribución e Historia Evolutiva. *Revista Peruana de Biología* 15(suppl. 1):7–29.

Prance, G.T., W. Balée, B.M. Boom, and R.I. Carneiro. 1987. Quantitative Ethnobotany and the Case for Conservation in Amazonia. *Conserv Biol* 1(4):296–310.

Rakotoarinivo, M. 2008. A Remarkable *Ravenea* from the Montane Forest of Andilamena, Madagascar. *Palms* 52(1):11–17.

Rakotoarinivo, M., T. Ranarivelo, and J. Dransfield. 2007. A New Species of *Beccariophoenix* from the High Plateau of Madagascar. *Palms* 51(2):63–75.

Reyes-Garcia, V., N. Martí, T. McDade, S. Tanner, and V. Vadez. 2007. Concepts and Methods in Studies Measuring Individual Ethnobotanical Knowledge. *J Ethnobiology* 27(2):182–203.

Ruelle, M.L., and K.A.S. Kassam. 2011. Diversity of Plant Knowledge as an Adaptive Asset: A Case Study with Standing Rock Elders. *Econ Bot* 65(3):295–307.

Sunderland, T. 2007. *Field Guide to the Rattan Palms of Africa.* Kew Publishing, Royal Botanic Gardens, Kew, Richmond.

Thomas, E., I. Vandebroek, and P. Van Damme. 2007. What Works in the Field? A Comparison of Different Interviewing Methods in Ethnobotany with Special Reference to the Use of Photographs. *Econ Bot* 61(4):376–384.

Trotter, R.T., and M.H. Logan. 1986. Informant Consensus: A New Approach for Identifying Potentially Effective Medicinal Plants. Pp. 91–112 in N.K. Etkin (ed.), *Plants in Indigenous Medicine and Diet.* Redgrave Publishing Company, New York.

Tuley, P. 1995. *The Palms of Africa.* Trendrine Press, St. Ives.

Vandebroek, I. 2010. The Dual Intracultural and Intercultural Relationship between Medicinal Plant Knowledge and Consensus. *Econ Bot* 64(4): 3–317.

Wallace, A.R. 1853. *Palm Trees of the Amazon and Their Uses.* Palm Publisher, London.

Wulijarni-Soitjipto, N. and S. Danimihardja (eds.). 1995. *Plant Resources of South-east Asia 6: Rattans.* Bogor, Indonesia.

## GENERAL INFORMATION ON THE COMMUNITY

Date: (dd.mm..yyyy) ______/______/ 2012 Country/Province/Community: 

Name of community leader (s): Position: 

### COMMUNITY SIZE (*)This information can be filled in detail when there is sufficient information from a community census

N° families: N° males (*) N° boys (*) (<18 yrs)

N° inhabitants: N° females (*) N° girls (*) (<18 yrs)

### HISTORY & ETHNIC BACKGROUND

Date of the community's foundation

Principal productive activities

**Ethnic background:** *Ethnic groups present in the community, their estimated proportion (% E) and estimated proportion of members from each ethnic group who speak their native languate (% S)*

| | | | | | | | | | | | | | |
|---|---|---|---|---|---|---|---|---|---|---|---|---|---|
| 1. | | % E | | % S | | | 6. | | % E | | % S | |
| 2. | | % E | | % S | | | 7. | | % E | | % S | |
| 3. | | % E | | % S | | | 8. | | % E | | % S | |
| 4. | | % E | | % S | | | 9. | | % E | | % S | |
| 5. | | % E | | % S | | | 10. | | % E | | % S | |

### BASIC SERVICES AVAILABLE WITHIN THE COMMUNITY

***Education***

Primary level: ____ years N° students primary level: Secondary level: ____ years N° students secondary level:

***Medical attention***

Health post: Hospital: Communal health worker: Traditional healer:

Other:

***Lighting source***

Overland line ☐ Public generator ☐
Personal generator ☐ Solar panel ☐
Oil lamp/candles ☐ Gas ☐
Other

***Drinking water source***

Public tap, NOT drinkable ☐ Public tap, drinkable ☐
Tap in house, NOT drinkable ☐ Tap in house, drinkable ☐
Well ☐ River/Stream ☐
Other

***Sewage system***

Yes ☐ No ☐ Other

***Sanitary system***

Toilets ☐ Latrine ☐ Septic tank ☐ Other

***Cooking fuel source***

Gas ☐ Firewood ☐ Coal ☐ Other

***Church/Mosque/Temple***

Yes ☐ No ☐ Other

***Communications services***

Radio ☐ Public phone ☐ Celular ☐ TV/DVD ☐ Other

Appendix 1. Socioeconomic and Historic Community Questionnaire. To be Filled with the Local Leader(s).

***Markets and/or supply sources***

Markets: Permanent [ ] Weekly [ ] Monthly [ ] Village shops: Yes [ ] No [ ] Number [ ]

Products sold: [ ]

***SOCIOECONOMIC INDICATORS***

***Type of tools in the community (and quantity)***

Tractor [ ] Plough [ ] Chainsaw [ ] Other [ ]

***Type of transportation in the community (and quantity)***

Truck [ ] Car [ ] Motorcycle [ ] Bicycle [ ] Canoe [ ] Outboard motor [ ] Other [ ]

***Type of animals in the community (and quantity)***

Cows [ ] Horses [ ] Mules [ ] Pigs [ ] Chickens [ ] Ducks [ ] Other [ ]

***Crops planted in the community***

[ ]

***ACCESSIBILITY*** **(Distance registered in Km for terrestial and in hours for fluvial transportation)**

Town/market/path: [ ] Transportation type: [ ] Distance: [ ]

Town/market/path: [ ] Transportation type: [ ] Distance: [ ]

Town/market/path: [ ] Transportation type: [ ] Distance: [ ]

***LAND***

Type of land ownership [ ]

Comunal area (ha) [ ] Populated area (ha) [ ] Forest area (ha) [ ]

Forest type found in the comunal area [ ]

***COMMUNITY MAP (with location of houses)***

***COMMUNITY AREA MAP (with vegetation types)***

Appendix 1 (cont.).

**COMMUNITY CENSUS**

Information provided by the interviewee:

- Number of families in each house. If more than one family is present, establish the relationships among them.
- Number of people in each family, gender and age of those above 18 years.

Other information registerd by the researcher:

- House census number
- Materials employed in the construction
    - Only local material
    - Only external material
    - Local and external material
- Palm species and local material used for:
    - Thatch
    - Walls
    - Floors
    - Structural material (posts, beams, etc.)
- House condition:
    - Good condition
    - Minor defects
    - Major defects
    - Poor condition

Appendix 2. Community census questionnaire. All questions to be filled out for each household visited.

**N° Informant:________**

| Date (dd.mm..yyyy) | ______/______/2012 | Community/N° house | ______________/______________ |
|---|---|---|---|

**PERSONAL PROFILE**

| Name: | | Communal post: | |
|---|---|---|---|

| Male | | Female | | Age | | Married | | Single | | Widow | | N° children | | N° people living in the house | |
|---|---|---|---|---|---|---|---|---|---|---|---|---|---|---|---|

| Place of birth | | Ethnic group | |
|---|---|---|---|
| Time of residence in this community | | | |

| Name of previous community(s) and time of residence | | | | | | |
|---|---|---|---|---|---|---|
| | 1. | | | 3. | | |
| | 2. | | | 4. | | |

| Years of schooling | | Years superior education | | Other | |
|---|---|---|---|---|---|

| Language(s) spoken (S), read (R) or written (W) | | | | | | | | | | | | | | | |
|---|---|---|---|---|---|---|---|---|---|---|---|---|---|---|---|
| | 1. | | S | | R | | W | | 3. | | S | | R | | W |
| | 2. | | S | | R | | W | | 4. | | S | | R | | W |

| Main occupation | |
|---|---|
| Main income source | |

**SOCIOECONOMIC INDICATORS**

***Lighting source***

| Overland line | | Public generator | |
|---|---|---|---|
| Personal generator | | Solar panel | |
| Oil lamp/candles | | Gas | |
| Other | | | |

***Drinking water source***

| Public tap, NOT drinkable | | Public tap, drinkable | |
|---|---|---|---|
| Tap in house , NOT drinkable | | Tap in house, drinkable | |
| Well | | River/Stream | |
| Other | | | |

***Cooking fuel source***

| Gas | | Firewood | | Coal | | Others | |
|---|---|---|---|---|---|---|---|

***Sanitary system***

| Toilet | | Latrine | | Septic tank | | Others | |
|---|---|---|---|---|---|---|---|

***Animals (and quantity)***

| Cows | | Horses | | Mules | | Pigs | | Chickens | | Ducks | | Other | |
|---|---|---|---|---|---|---|---|---|---|---|---|---|---|

***Crops or plantations (and size in hectares or square meters)***

| |
|---|
| |

Appendix 3. Informant socioeconomic information.

***Tools and utensils (and quantity)***

Tractor [ ] Plough [ ] Shovel [ ] Pick ax [ ] Machet [ ] Planting stick [ ] Ax [ ] Chainsaw [ ]

Shotgun [ ] Bow/arrow [ ] Blowgun [ ] Fish-hooks [ ] Fishing-rod [ ] Fishing-net [ ] Harpoon [ ]

***Transportation owned by family (and quantity)***

Truck [ ] Car [ ] Motorcycle [ ] Bicycle [ ] Canoe [ ] Outboard motor [ ]

***Number of markets and supply centers attended regularly (city, town, other community)***

Place [ ] Frequency (weekly, monthly, etc.) [ ]

Products bought [ ]

Products sold [ ]

Place [ ] Frequency (weekly, monthly, etc.) [ ]

Products bought [ ]

Products sold [ ]

## USE INFORMATION

***Perception of species quality for certain use categories***

| | | | | | |
|---|---|---|---|---|---|
| **Human food** | 1. | 2. | 3. | 4. | 5. |
| **Oils** | 1. | 2. | 3. | 4. | 5. |
| **Thatching** | 1. | 2. | 3. | 4. | 5. |
| **Walls** | 1. | 2. | 3. | 4. | 5. |
| **Framework** | 1. | 2. | 3. | 4. | 5. |
| **Local markets** | 1. | 2. | 3. | 4. | 5. |
| **Regional markets** | 1. | 2. | 3. | 4. | 5. |

***Of the species known, which one is the most important? Why?***

[ ]

***Knowledge Transmission***

***Is the knowledge on palm uses transmitted within the community? Why?*** Yes No

[ ]

Appendix 3 (cont.).

**PALM USE QUESTIONARY – GENERAL INFORMANT**

1. Vernacular (local) name(s) of the palm

   Language of vernacular name and meaning of the name

2. Are there different varieties of the species? How are they differentiated?

3. What are the uses for this species? For reference use the categories and subcategories of use refer to Annex 5, for plant parts used to Annex 6.

   Which products are obtained?

   What is the palm part used for each case? (specify when possible the local names for the products)

   Which is the frequency of use? Is it a: **a)** Actual or **b)** Past use? If it is a past use, when was the last time it was used?

   Where is the resource collected? **a)** Agroforestry systems, **b)** Garden/courtyard **c)** Primary forest **d)** Plantations **e)** Palm stand **f)** Grasslands **g)** Secondary forest

   What are the techniques of harvesting? **a)** Destructive: necessary felling **b)** Destructive: unnecessary felling **c)** Non-destructive: climbing **d)** Non-destructive: harvest of cespitose palms (felling few or no trunks) **e)** Non-destructive: direct harvest of low palms (also juvenile, sub adults) or acaulescents **f)** Non-destructive: harvest (collection) on the ground **g)** Non-destructive: but without information

4. Abundance of the species in the past 10 years: **a)** Same **b)** Increased **c)** Decreased

5. Use of the species in the past 10 years: **a)** Same **b)** Increased **c)** Decreased

6. Commercialization: Type of product / Market type: (L) local; (R) regional; (N) national; (E) exportation / Frequency (daily, weekly, monthly, etc.) / Quantity / Price

Appendix 4. Palm use questionnaire for general and expert informants. Options in the questions techniques of harvesting and management systems are based on Bernal et al. 2011.

**PALM USE QUESTIONARY – EXPERT INFORMANT**

1. Vernacular (local) name(s) of the palm:

   Language of vernacular name and meaning of the name:

2. Are there different varieties of the species? How are they differentiated?

3. What are the uses for this species? For reference use the categories and subcategories of use refer to Annex 5, for plant parts used to Annex 6.

   Which products are obtained?

   What is the palm part used for each case? (Specify when possible the local names for the products)

   When is the resource collected? according to age of the plant: **a)** Seedling **b)** Juvenile **c)** Sub adult **d)** Adult **e)** Indistinct

   What are the characteristics of the processing? **a)** Without processing, direct use **b)** With processing (describe). For medicinal use, indicate information on preparation, administration and contraindications (if any)

   Which is the frequency of use? **a)** Actual or **b)** Past use? If it is a past use, when was the last time it was used?

   Where is the resource collected? **a)** Agroforestry systems, **b)** Garden/courtyard **c)** Primary forest **d)** Plantations **e)** Palm stand **f)** Grasslands **g)** Secondary forest

   What are the techniques of harvesting? **a)** Destructive: necessary felling **b)** Destructive: unnecessary felling **c)** Non-destructive: climbing **d)** Non-destructive: harvest of cespitose palms (felling few or no trunks) **e)** Non-destructive: direct harvest of low palms (also juvenile, sub adults) or acaulescents **f)** Non-destructive: harvest (collection) on the ground **g)** Non-destructive: but without information

   Is there a management system? **a)** Cultivation **b)** Enrichment areas of harvest with dispersion of seeds or seedlings planting **c)** Fertilization **d)** Use of fire (to increase the presence of palms) **e)** Rotating the harvest area **f)** Leaving the palms when the forest is cut **g)** Selective harvest by age, size and/or sex **h)** Seasonal restriction (moon phases, phenology or climate) **i)** Individuals or areas left as seed banks **j)** Pruning

Appendix 4 (cont.).

**k)** Transplanting **l)** Clearing other shrubs, trees and lianas (eliminate competition) **m)** They have a formal study (e.g. Management Plan)

4. How is it distributed in the vicinity of the community? **a)** Abundant **b)** Moderate (common) **c)** Rare **d)** Cultivated **e)** Other (describe)

5. Abundance of the species in the past 10 years: **a)** Same **b)** Increased **c)** Decreased

6. Use of the species in the past 10 years: **a)** Same **b)** Increased **c)** Decreased

7. Commercialization: Type of product / Market type: (L) local; (R) regional; (N) national; (E) exportation / Frequency (daily, weekly, monthly, etc.) / Quantity / Price

Appendix 4 (cont.).

| Use category | Use subcategory | Description |
|---|---|---|
| Animal Food | Fish Bait | Bait for fishing |
| | Fodder | Food for domestic animals |
| | Wildlife Attractant | Palms that provide food for mammals and whose location constitutes preferential areas for hunting |
| Construction | Bridges | Materials to bridge watercourses |
| | Houses | Houses and other constructions such as temporary camps, animal yards |
| | Thatch | House thatching and other constructions |
| | Transportation | Canoes, rafts, oars and other materials for sealing |
| | Other | Uses not classifiable within the previous subcategories, for example stems used as posts for telephone lines and gutters to transport water |
| Cultural Uses | Clothes and Accessories | Articles of clothing and accessories such as hats |
| | Cosmetic | Beauty products, including perfumes, oils, shampoo, and other hair care products |
| | Dyes | Dyeing of diverse materials (vegetables) and as body paint |
| | Personal Adornment | Necklaces, bracelets, earrings, armbands, pectorals, anklets |
| | Recreational | Musical instruments, toys, ashes as additives to the consumption of tobacco and coca leaves |
| | Ritual | Uses related to myth-religious aspects, including festivals and feasts, construction of coffins, to drive away feared animals, sorcery |
| | Other | Uses not classifiable under the previous subcategories |
| Environmental Uses | Agroforestry | Palms that are part of agroforestry systems with different management degrees |
| | Fences | Delimitation of properties, barriers |
| | Ornamental | Palms cultivated for ornamental purposes |
| | Soil Improvers | Fertilizers, edaphic protectors and agents against soil erosion |
| Fuel | Firewood | Wood to make fire |
| | Fire Starter | Combustion starters |
| | Lighting | Lamps, torches and candles |
| | Other | Uses not classifiable within the previous subcategories, for example for waterproofing canoes |
| Human Food | Beverages | Elaboration of unfermented or fermented drinks |
| | Food | Edible, generally with little preparation |
| | Food Additives | Ingredients used in the preparation and processing of foods |
| | Oils | Edible fats |
| Medicinal and Veterinary | Blood and Cardio-vascular System | Anemia, cardiovascular problems and ailments, cardiac diseases, varicose veins, hypertension, hypotension, haemorrhoids |

Appendix 5. Description of the use categories and subcategories under which ethnobotanical information is categorized for palm uses during interviews. Use categories and subcategories are based in Cook (1995) with adaptations for tropical regions. This has also been published in Macía et al. (2011).

| | | |
|---|---|---|
| | Cultural Diseases and Disorders | Ailments or disorders of magic-religious origin recognized by a specific culture, like mal aire ('bad air'), arrebato ('outburst'), susto, huaraña |
| | Dental Health | Caries, tooth pains, fillings, dental hygiene |
| | Digestive System | Carminative, colics, flatulence, emetic, indigestion, purgative, gastric or intestinal ulcers, diarrhea, laxatives, liver and vesicular disorders, hepatitis |
| | Endocrine System | Diabetes |
| | General Ailments with Unspecific Symptoms | General ailments like body pains, general discomfort, weakness, headache, fever |
| | Infections and Infestations | Malaria, leishmaniasis, measles, antihelminthic, louse, fleas, chiggers, scabies |
| | Metabolic System and Nutrition | Obesity, weight loss |
| | Muscular-Skeletal System | Rheumatism, twists, fractures, sciatic, lumbalgia |
| | Nervous System and Mental Health | Migraine, mental disorders, epilepsy, paralysis, nervous disorders |
| | Poisoning | Snakebites, scorpion stings, rays, spiders, insects |
| | Pregnancy, Birth and Puerperium | Gestation, haemorrhage, childbirth, postnatal, lactation, abortive, postpartum |
| | Reproductive System and Reproductive Health | Menstruation, fertility, venereal diseases, prostrate, impotence, menopause, aphrodisiacs, contraceptives |
| | Respiratory System | Flu, cold, loss of voice, bronchitis, pneumonia, expectorant, cough |
| | Sensory System | Eye infections, cataracts, loss of sight or smell, deafness, ear infection |
| | Skin and Subcutaneous Tissue | Acne, boils, eczemas, burns, extraction of spines stuck on the skin |
| | Urinary System | Diuretic, kidney stones, urinary incontinence, urinary infections, cystitis |
| | Veterinary | Treatment of diseases or ailments for domestic animals |
| | Not Specified | Medicinal use or with pharmacological properties, but with insufficient information to assign to one of the described subcategories |
| | Other | Uses not classifiable within the previous subcategories, for example tumours, cancer, anaesthetic |
| Toxic | Fishing | Fish poison |
| | Hunting | Poison for hunting |
| Utensils and Tools | Domestic Utensils | Baskets, fans, hammocks, bags, domestic furniture, air freshener |
| | Hunting and Fishing Tools | Bows, arrows, blowpipes, harpoons, fishing nets, hunting traps |
| | Labour Tools | Agricultural or domestic tools like spinners, machetes and lubricants of these materials |
| | Rope | Manufacturing of ropes and moorings |
| | Wrappers | Wrappers for materials and foods |
| | Other | Uses not classifiable within the previous subcategories, for example insect repellents |
| Other Uses | Miscellaneous | Uses not classifiable within the previous categories. Indirect use of palms: insect larvae feeding on rotting stems used as food, medicine or bait |

Appendix 5 (cont.).

ANNEX 6. Palm parts mentioned in the description of uses by the informants during interviews.

- Complete plant
- Root: fulcreous, adventitious, subterranean
- Stem
- Spines (stem)
- Leaf
- Leaf sheath
- Petiole
- Lamina (leaf)
- Rachis (leaf)
- New leaf (unnopened)
- Palm heart
- Peduncular bract
- Inflorescences
- Flowers
- Infrutescence
- Fruits: exo, meso, endocarp
- Endosperm

Appendix 6. Palm parts mentioned in the description of uses by the informants during interviews.

| Species | Priority of registering use information | Description |
|---|---|---|
| *Astrocaryum* sect. *huicungo*<br>*Astrocaryum chambira* Burret<br>*Attalea butyracea* (Mutis ex L.f) Wess. Boer<br>*Astrocaryum standleyanum* L.H. Bailey<br>*Bactris gasipaes* Kunth<br>*Ceroxylon* spp.<br>*Euterpe precatoria* Mart.<br>*Geonoma deversa* (Poit.) Kunth<br>*Geonoma macrostachys* Mart.<br>*Iriartea deltoidea* Ruiz & Pav.<br>*Mauritia flexuosa* L.f.<br>*Oenocarpus bataua* Mart.<br>*Phytelephas macrocarpa* Ruiz & Pav.<br>*Socratea* exorrhiza (Mart.) H. Wendl. | High | These species will be asked to all informants (experts and general) in all study sites. |
| *Attalea maripa* (Aubl.) Mart.<br>*Bactris acanthocarpa* Mart.<br>*Bactris concinna* Mart.<br>*Bactris maraja* Mart.<br>*Chamaedorea pinnatifrons* (Jacq.) Oerst.<br>*Cocos nucifera* L.<br>*Desmoncus mitis* Mart.<br>*Desmoncus polyacanthos* Mart.<br>*Geonoma maxima* (Poit.) Kunth<br>*Geonoma stricta* (Poit.) Kunth<br>*Mauritiella armata* (Mart.) Burret<br>*Oenocarpus mapora* H. Karst. | Medium | These species will be asked to all informants only if they are found in the forest or in the community. |

Appendix 7. List of common useful palms species growing in Amazonian forests of Colombia, Ecuador, Perú y Bolivia (based on Macía et al. 2011) from which ethnobotanical information will be registered with all informants in the areas the species exist.

CARIBBEAN:

**Regional:**
Henderson, A., Galeano, G. & R. Bernal. 1997. Field Guide to the Palms of the Americas. Princeton University Press, Princeton. 363 p.

Read, R.W. 1979. Palmae. Pp. 320—368 in Howard, R.A. (ed.), Flora of the Lesser Antilles-Monocotyledoneae. Brooke Thomson-Mills, Massachusetts.

**Antigua and Barbuda:** refer to Regional references

**Bahamas:** refer to Regional references

**Barbados:** refer to Regional references

**Cuba:**
Leiva S., A. 2001. Cuba y sus palmas. Instituto Cubano del Libro. Editorial Gente Nueva, La Habana. 73 p.

Moya L., C.E. & A.T. Leiva S. 2000. Checklist of the palms of Cuba, with notes on their ecology, distribution and conservation. Palms 44 (2): 69--84.

**Dominica:** refer to Regional references
James, A. 2009. Notes on the uses of Dominica's Native Palms. Palms 53(2): 61--67.

**Dominican Republic:**
Hoppe, J. 1998. Palms of the Dominican Republic. Fundación Manuel de Jesus Tavares Portes, Santo Domingo. 106 p.

**Grenada:** refer to Regional references

**Guadeloupe & Martinique:** refer to Regional references
Fournet, J. 1978. Flore illustrée des phanerogames de Guadeloupe et de Martinique. I.N.R.A., Paris. 2538 p.

**Haiti:** refer to Regional references

**Jamaica:** refer to Regional references

**Puerto Rico:** refer to Regional references

**Saint Kitts and Nevis:** refer to Regional references

**Saint Lucia:** refer to Regional references

**Saint Vincent and the Grenadines:** refer to Regional references

**Trinidad & Tobago:** refer to Regional references
Comeau, P.L., Comeau, Y.S. & W. Johnson. 2003. The Palm Book of Trinidad and Tobago, including the Lesser Antilles. The International Palm Society. 108 p.

MESOAMERICA:

**Regional:**
Henderson, A., Galeano, G. & R. Bernal. 1997. Field Guide to the Palms of the Americas. Princeton University Press, Princeton. 363 p.

Appendix 8. Tropical palm field guides and checklists. References are arranged by geographical region. For each region listed, first is the general reference, followed by the remaining guides sorted alphabetically by countries. Country references are sorted chronologically from most recent to the oldest.

**Belize:**

Brewer, S.W. 1999. The palms of Belize: Species richness and a key based on vegetative characters. Palms 43(3): 109-113.

**Costa Rica:**

Grayum, M.H. 2003. Arecaceae. Pp. 201--293 in Hammel, B.H., Grayum, M. H., Herrera, C. & N. Zamora (eds.), Manual de plantas de Costa Rica. Vol. II. Gimnospermas y monocotiledóneas (Agavaceae-Musaceae). Monogr. Syst. Bot. Missouri Bot. Gard. 92.

**El Salvador:** refer to Regional references

**Guatemala:** refer to Regional references

Standley, P. & J. Steyermark. 1958. Palmae. Flora of Guatemala. Fieldiana Botany 24:196--299.

**Honduras:** refer to Regional references

**Mexico:**

Quero H.J. & J.S. Flores. 2004. Arecaceae de la Península de Yucatán. Etnoflora Yucatanense 23:1–111.

Quero, H.J. 1992. Las Palmas silvestres de la península de Yucatán. Universidad Nacional Autónoma de México, Mexico City. 63 p.

**Nicaragua:**

Read, R.W., Henderson, A., Ulloa Ulloa, C. & R. Evans. 2001. Arecaceae. Pp. 192--229 in Stevens, W.D., Ulloa Ulloa, C., Pool, A. & O.M. Montiel (eds.), Flora de Nicaragua Vol. 1. (Acanthaceae – Euphorbiaceae). Missouri Botanical Garden Press, St. Louis.

**Panama:** refer to Regional references

Correa, M.D., Galdames, C. & M.S. de Stapf. 2004. Catálogo de las plantas vasculares de Panamá. Quebecor World Bogotá, S.A., Bogotá.

SOUTH AMERICA

**Regional:**

Henderson, A., Galeano, G. & R. Bernal. 1997. Field Guide to the Palms of the Americas. Princeton University Press, Princeton. 363 p.

Henderson, A. 1995. The palms of the Amazon. Oxford University Press, New York. 388 p.

**Bolivia:**

Moraes, M. 2004. Flora de palmeras de Bolivia. Herbario Nacional de Bolivia, La Paz. 262 p.

**Brazil:**

Lorenzi, H., Noblick, L., Kahn, F. & E. Ferreira 2010. Flora Brasileira (Arecaceae). Editora Plantarum, Nova Odessa. 384 p.

Miranda, I. P. A. & A. Rabelo. 2008. Guia de identificaçao das palmeiras de Porto Trombetas – PA. INPA, Manaus. 364 p.

Lorenzi, H., Souza, H.M., Costa, J. T. M., Ferreira, E. 2004. Palmeiras brasileiras e exoticas cultivadas. Vol. 02. Editora Plantarum, Nova Odessa. 416 p.

**Chile:**

Grau, J. 2006. Palms of Chile: a detailed investigation of the two endemic palms and a review of introduced species. Ediciones OIKOS, Santiago de Chile. 203 p.

**Colombia:**

Galeano, G. & R. Bernal. 2010. Palmas de Colombia. Guía de Campo. Universidad Nacional de Colombia. Instituto de Ciencias Naturales, Bogotá. 688 p.

**Ecuador:**

Appendix 8 (cont.).

Borchsenius, F., Borgtoft Pedersen, H. & H. Balslev. 1998. Manual of the palms of Ecuador. University of Aarhus, Aarhus. 217 p.

**French Guiana:** refer to Regional references

**Guyana:** refer to Regional references

**Paraguay:**
Hahn, W.H. 1990. A synopsis of the Palmae of Paraguay. MSc thesis, Cornell University. 454 p.

**Peru:** refer to Regional references
Kahn, F. & F. Moussa. 1994. Las palmeras de Perú. Instituto Francés de Estudios Andinos-IFEA. Lima. 180 p.

**Surinam:** refer to Regional references
Wessels Boer, J.G. 1965. The indigenous palms of Suriname. E.J. Brill, Leiden. 173 p.

**Venezuela:**
Hoyos, J & A. Braun. 2001. Palmas en Venezuela – Autóctonas y exóticas. Sociedad de Ciencias Naturales Lasalle, Caracas.

Braun. A. & F. Deslascio Chitty. 1987. Palmas autóctonas de Venezuela. LITOPAR C.A. Caracas. 156 p.

AFRICA & MADAGASCAR

**Regional:**
Dransfield, J. 2010. Arecaceae in Timberlake, J. (ed.), Flora Zambesiaca 13 (2). Kew Publishing, Royal Botanic Gardens, Kew, Richmond.

Sunderland, T. 2007. Field Guide to the Rattan Palms of Africa. Kew Publishing, Royal Botanic Gardens, Kew, Richmond. 66 p.

Tuley, P. 1995. The Palms of Africa. Trendrine Press, St. Ives. 189 p.

Dransfield, J. 1986. Palmae. Pp. 1--60 in Polhill, R.M. (ed.), Flora of Tropical East Africa. A.A. Balkema, Rotterdam. (12 species)

Russell, T.A. 1968. Palmae. Pp. 159—169 in Hutchinson, J. (ed.), Flora of west tropical Africa 3(1). Royal Botanic Gardens Kew, Crown Agents, London.

Moore Jr., H. E. & L.J. Gueho. 1984. Palmiers (34 pp.) in Bosser, J., Cadet, T., Gueho, L. J. & W. Marais (eds.), Flore des Mascareignes. The Sugar Industry Research Institute, Mauritius, ORSTOM, Paris & The Royal Botanic Gardens, Kew.

**Benin:**
Aké Assi, L., van der Maesen, L.J.G. & J. Dransfield. 2006. Arecaceae. Pp. 50—62 in Akoegninou, A., van der Burg, W.J. & L.J.G. van der Maesen (eds), Flore Analytique du Bénin. Backhuys Publishers, Cotonou and Wageningen..

**Equatorial Guinea:**
Sunderland, T.C.H. 1998. The rattans of Rio Muni, Equatorial Guinea: utilisation, biology and distribution; a report for the Projecto Convservación y Utilización Forestales de Guinea Ecuatorial (CUREF) and the Ministerio de Pesca y Forestal, Guinea Ecuatorial. Kew: African Rattan Research Programme. 28 p.

**Madagascar:**
Dransfield, J., H. Beentje, Britt, A., Ranarivelo, T. & J. Razafitsalama. 2006. Field guide to the palms of Madagascar. Kew Publishing, Royal Botanic Gardens, Kew, Richmond. 176 p. (Published in two editions – English and Malagasy)

Dransfield, J. & H. Beentje. 1995. The Palms of Madagascar. Kew Publishing, Royal Botanic Gardens, Kew, Richmond. 475 p.

**Seychelles:**

Appendix 8 (cont.).

Matatiken D. & D. Dogley. 2006. Guide to the Endemic Palms and Screw Pines of the Seychelles Granitic Islands. Plant Conservation Action Group. 45 p.

**South Africa:**

Pooley, E.S. 1989. Palms of Southern Africa. Veld and Flora, magazine of the Botanical Society of SA.

Wicht, H. 1969. The indigenous palms of southern Africa. Howard Timmins, Cape Town. 62 p.

ASIA

**Regional:**

Henderson, A. 2009. Palms of Southern Asia. Princeton University Press. 264 p.

Dransfield, J. & M. Patel. 2005. Rattans of Borneo: an interactive key. CD Rom. Kew Publishing, Royal Botanic Gardens, Kew, Richmond.

Dransfield, J. & N. Manokaran (eds.). 1993. Plant resources of South-East Asia no. 6: Rattans. Pudoc Scientific Publishers, Wageningen. 137 p.

**Bangladesh:**

Alam, M.K. 1990. Rattans of Bangladesh. Bangladesh Forest Research Institute, Chittagong. 34 p.

**Bhutan**

Noltie, H.J. 1994. Arecaceae. Flora of Bhutan, Vol.3 Part 1. Royal Botanic Garden, Edinburugh.

**Brunei:**

Dransfield, J. 1997. The rattans of Brunei Darussalam. Ministry of Industry and Primary Resources, Brunei Darussalam. 217 p.

**Cambodia:**

Khou E. 2008. A field guide of the rattans of Cambodia. World Wildlife Fund, Cambodia. 71 p.

**China:**

Pei, S., Chen, S., Lixiu, G., Dransfield, J. & A. Henderson. 2010. Arecaceae. Pp. 133--157 in Wu, Z. Y., P. H. Raven & D. Y. Hong (eds.). Flora of China. Vol. 23. Science Press, Beijing, and Missouri Botanical Garden Press, St. Louis.

**East Timor:** refer to Indonesia

**India:**

Renuka, C. 2008. Field identification key for Indian palms. Kerala Forest Research Institute. Peechi, India. CD-ROM.

Renuka, C. & K.M. Bhat. 2002. Commercial rattans of Kerala. Kerala Forest Research Institute. Peechi. 25 p.

Renuka, C. 2001. Field identification key for rattans of Kerala. Kerala Forest Research Institute, Peechi. 33 p.

Renuka, C. 1999. Palms of Kerala. Kerala Forest Research Institute. Peechi. 72 p.

Renuka, C. 1995. A manual of the rattans of Andaman and Nicobar islands. Kerala Forest Research Institute. Peechi. 72 p.

Basu, S.K. and Chakraverty, R.K. 1994. *A Manual of cultivated Palms in India.* Botanical Survey of India, Kolkata.

Lakshmana, A.C. 1993. Rattans of South India. Evergreen Publishers, Bangalore. 180 p.

Basu, S.K. 1992. Rattans (canes) in India: a monographic revision. Kepong, Kuala Lumpur: Rattan Information Centre. 141 p.

Renuka, C. 1992. Rattans of the Western Ghats: a taxonomic manual. Kerala Forest Research Institute. Peechi. 61 p.

**Indonesia:**

Appendix 8 (cont.).

Powling, A. 2009. The Palms of Buton, Indonesia, an island in Wallacea. Palms 53(2): 84--91.

Keßler, P.J.A., Bos, M.M., Sierra-Daza, S.E.C., Kop, A., Willemse, L.P.M., Pitopang, R. & S.R. Gradstein. 2002. Checklist of woody plants of Sulawesi, Indonesia. Blumea Supplement 14. Nationaal Herbarium Nederland, Universiteit Leiden branch, Leiden.

Visser, M.B.H. 1991. 100 Macam palem di Indonesia. Ministry of Education and Sciences, the Netherlands. 57 p.

Sastrapradja, S., Mogea, J.P., Sangat, H.M. & J.J. Afriastini. 1987. Palms of Indonesia: The English Translation of Palem Indonesia. The Palm and Cycad Societies of Australia, Milton. 39 p.

Sastrapradja, S., Mogea, J.P., Sangat, H.M. & J.J. Afriastini. 1978. Palem Indonesia. Lembaga Biologi Nasional, Bogor. 120 p.

House, A.P.N. 1983. The use of palms by man on Siberut Island, Indonesia. Principes 27(1): 12--17.

Dransfield, J. 1974. Notes on the palm flora of central Sumatra. Reinwardtia 8(4): 519-531.

**Iraq:**

Dransfield, J. 1985. Palmae. Pp. 260—267 in Townsend, C.C. & E. Guest (ed.). Flora of Iraq 8. Ministry of Agriculture, Baghdad and Bentham-Moxon Trust.

**Japan:** refer to Regional references

**Lao:**

Evans, T.D. Sengdala, K., O. V. Viengkham & B. Thammavong. 2001. Field Guide to the Rattans of LAO PDR. Kew Publishing, Royal Botanic Gardens, Kew, Richmond. 96 p.

**Malaya:**

Whitmore, T.C. 1998. Palms of Malaya. White Lotus Co. Ltd., Bangkok. 136 p.

**Malaysia:**

Pearce, K.G. 2003. The palms of Kubah National Park, Matang, Kuching Division, Sarawak. Final Report WWF 3325. 95 p.

Dransfield, J. 1992. The rattans of Sarawak. Kew Publishing, Royal Botanic Gardens, Kew, Richmond. 233 p.

Dransfield, J. 1984. The rattans of Sabah. Forest Dept., Sabah. 182 p.

Dransfield, J. 1984. The palm flora of the G. Mulu National Park. Pp. 41--75 in Parris, B.S. (ed.), Studies on the flora of G. Mulu National Park, Sarawak. Kuching, Sarawak.

Dransfield, J. 1979. A manual of the rattans of the Malay Peninsula. Forest Department, Minsitry of Primary Industries, Kuala Lumpur. 270 p.

**Myanmar:** refer to Regional references

**Nepal:**

Amatya, S.M. 1997. The rattans of Nepal. IUCN, Kathmandu. 26 p.

**New Guinea:**

Baker, W.J. & J. Dransfield. 2006. Field guide to the palms of New Guinea. Kew Publishing, Royal Botanic Gardens, Kew, Richmond. 108 p. (Published in two editions – English and Indonesian)

Barfod, A.S., Banka, R. & J. L. Dowe. 2001. Field guide to the palms of Papua New Guinea. 79 p.

Ferraro, M. 1997. A checklist of Palmae for New Guinea. Palms & Cycads (Australia) 55/56: 2—39.

**Pakistan:**

Malik, K.A. 1984. Palmae. Pp. 1--33 in Nasir, E. & S.I. Ali (eds.), Flora of Pakistan 153. University of Karachi, Islamabad.

Appendix 8 (cont.).

**Philippines:**
Baja-Lapis, A. 2010. A field guide to Philippine rattans. 214 p.

Forero, E.S. 1990. A preliminary analysis of the palm flora of the Philippine Islands. Principes 34(1): 28—45.

Guzman, E.D. de & E.S. Fernando. 1986. Philippine palms. Pp. 145--233 in Santos, J.V., Guzman, E.D. de & E.S. Fernando (eds.), Guide to the Philippine Flora and Fauna, Volume 4. Natural Resource Management Center, Philippines.

Beccari, O. 1919. The Palms of the Philippine Islands. Philipp. J. Sci. (Bot.) 14(3): 295-362.

**Singapore:** refer to Malaya & Malaysia

**Sri Lanka:**
Zoysa, N. de. 1994. Rattans of Sri Lanka: An illustrated field guide. Sri Lanka Forestry Department, Battaramulla. 82 p.

**Taiwan:**
Liao, J-C. 1994. Illustrations of the family Palmae in Taiwan. Department of Forestry, College of Agriculture, National Taiwan University, Taipei, Taiwan, Republic of China. 120 p.

**Thailand:**
Dransfield, J., Barfod, A.S. & R. Pongsattayapipat. 2004. A preliminary checklist to Thai Palms. Thai. For. Bull. (Bot.) 32: 32—72.

Hodel, D.R (ed.). 1998. The Palms and Cycads of Thailand. Allen Press, Lawrence. 190 p.

**Vietnam:**
Ho, P. H. 1993. Câycó Viêtnam. An illustrated flora of Vietnam 3(2). Published by the author.

AUSTRALIA & THE PACIFIC:

**Regional:**
Dowe, J.L. 1989. Palms of the South-West Pacific. Palm & Cycad Scoeity of Australia, Milton. 208 p.

Moore, H.E. & F.R. Fosberg 1956. The Palms of Micronesia and the Bonin Island. Ocasional Papers on the Kinds of Plants. Vol. 8, Fasc. 6. Gentes Herbarium, the Bailey Hortorium of the New York State Collage of Agriculture, New York. 733 p.

**Australia:**
Dowe, J.L. 2010. Australian Palms: Biogeography, Ecology and Systematics. CSIRO Publishing, Collingwood, Victoria, Australia. 304 p.

Cronin, L. 2003. Australian Palms, Ferns, Cycads and Pandans (2nd edition). Jon Carpenter Publishing, Chipping Norton. 190 p.

Tucker, R. 1988. Palms of subequatorial Queensland. Palm & Cycad Society of Australia, Milton. 100 p.

**Bismarck Archipelago:**
Essig, F.B. 1995. A Checklist and Analysis of the Palms of the Bismarck Archipelago. Principes 39 (3): 123--129.

**Fiji:**
Watling, D. 2005. Palms of the Fiji Islands. Environmental Consultants Fiji, Suva. 192 p.

**New Caledonia:**
Hodel, D.R. & J.C. Pintaud. 1998. The Palms of New Caledonia – Les palmiers de Nouvelle-Calédonie. Allen Press, Lawrence, Kansas. 120 p. (English & French)

**Palau:**
Hillmann-Kitalong, A., DeMeo, R.A., & T. Hola. 2008. Native Trees of Palau, A Field Guide. Self published by authors. 236 p.

Appendix 8 (cont.).

**Pohnpei:**

Herrera, K., Lorence, D.H., Flynn, T. & M.J. Balick. 2010. Checklist of the Vascular Plants of Pohnpei, Federated states of Micronesia with Local Names and Uses. *Allertonia* 10(1): 1-192.

**Solomon Islands:**

Dennis, G. 1989. Palms of the Solomon Islands. The Publication Fund, Palm & Cycad Societies of Australia, Milton. 60 p.

**Samoa**

Whistler, W.A. 1992. The palms of Samoa. Mooreana 2(3): 24--29.

**South West Pacific:**

Dowe, J.L. 1989. Palms of the south-west Pacific: their origin, distribution, and description. The Publication Fund, Palm & Cycad Societies of Australia, Milton. 198 p.

**Vanuatu:**

Dowe, J.L. & P. Cabalion. 1996. A taxonomic account of Arecaceae in Vanuatu, with descriptions of three new species. Australian Systematic Botany 9 (1): 1--60.

**Internet PALM references:**

www.fp7-palms.org

www.palmweb.org

www.palmbase.org

www.eunops.org

Appendix 8 (cont.).

# Biodiversity as a Resource: Plant Use, Land Use, and Nature Conservation in Shuar, Saraguro, and Mestizo Communities in South-Eastern Ecuador

Andrés Gerique[1]

The tropical premontane and montane rainforests of south-eastern Ecuador constitute one of the most important hotspots of plant diversity. However, the Andes of Ecuador are known as the region with the highest deforestation rate in South America. The experience in international nature conservation during the past decades has shown that, to be effective, biodiversity conservation in inhabited areas should provide welfare through the use of forest for local peoples who coexist with it. Based on the foregoing, this study pursued the following targets: (1) to document the ethnobotany of the different ethnic groups who live in the area of study - namely the Shuar, the Saraguros, and the mestizos, (2) to analyze actual land use, and (3), to evaluate the suitability of the implementation of agroforestry systems, ecotourism, payments for environmental services, and bioprospecting as instruments in line with the "protection by use" concept.

## 1. Introduction

Among terrestrial ecosystems, tropical rainforests host the richest biological diversity (Barthlott et al. 2007, Myers et al. 2000). The conservation of forests became a matter of international interest during the early 1970s (Wittemyer et al. 2008, Bruner et al. 2001, Sayer et al. 2000, Chazdon 1998). Nature reserves were created all around the world and an intense debate over conservation strategies between advocates of people-free parks (who argued that the presence of permanent communities within reserves is antithetical to conservation) and community-based conservationists followed. The latter maintained that protected areas need human occupants to protect them and that conservation will only be legitimate if local communities benefit from conservation (cf. Durand and Lazos 2008, Terborgh and Peres 2002). Sinclair et al. (2000), however, pointed out that both types of strategies are not alternatives, they are complementary. Strictly protected areas should be established within larger managed landscapes occupied by people (Colchester 2000). On this point, the fact that less than two decades ago around 80% of the protected areas in South America was already inhabited by indigenous peoples (Colchester 1996) and that the density of people in biodiversity hotspots is 73 per km², compared with a global average of 42 per km² (Cincotta 2000), cannot be ignored. To protect biodiversity, it is more logical to work with these peoples than "to cast them into the role of environmental villains and expel them from their homelands" (Colchester 2000). Consequently, the conservation emphasis shifted from protection and exclusion to the sustainable use of biodiversity and to prevention (Van Schaik and Rijksen 2002). The new conservation strategy is to provide welfare through the use of forest biodiversity for local peoples who coexist with it reflecting the concept of "protection by use" instead of "protection from use" under the philosophy coined by Janzen, a University of

[1] Institute of Geography, University of Erlangen-Nuremberg, Kochstr. 4/4, 91054 Erlangen, Germany. Email: andres.gerique-zipfel@geographie.uni-erlangen.de.

Pennsylvania biologist (cf. Pohle and Gerique 2006, Daily and Ellison 2002).

Moreover, as highlighted in the World Development Report 2007 (World Bank 2007), dependence of indigenous and local peoples on biological has generated enormous bodies of traditional ecological knowledge and traditional forest utilization systems that are thought to be sustainable. To describe this knowledge is an urgent task, as there is consensus about the fact that the rapid integration of indigenous and local people into the market economy is leading to a loss of the traditional plant knowledge and to a more intense exploitation of the forest resources, situating under threat their long-term sustainability (cf. Gray et al. 2008).

Agroforestry and community-based ecotourism are classic "protection by use" tools for biodiversity conservation in human-dominated landscapes (cf. Bhagwat et al. 2008, Robertson and Wunder 2005). In recent times, new tools such as Payments for Environmental Services (PES) have been developed. PES are economic incentives for local inhabitants not only for incorporating sustainable practices into resource management but also as a compensation for conserving biodiversity (Forest Trends et al. 2008). In addition, benefits from bioprospecting through in-country research on local biodiversity have recently been proposed as a further resource for conservation (cf. Kursar et al. 2007).

## 2. Forests, Conservation, and People in South-eastern Ecuador

According to different authors (Brehm et al. 2008, Barthlott et al. 2007, Myers et al. 2000), the tropical mountain rainforests of the eastern Andean slopes and the tropical premontane rainforests of the sub-Andean ranges in south-eastern Ecuador (Loja and Zamora Chinchipe Provinces) constitute one of the most important hotspots of plant diversity worldwide. However, the Andes of Ecuador are known as the region with the highest deforestation rate in South America (Günter et al. 2009). Regional biodiversity has come under a great pressure from the expansion of agriculture, infrastructure, logging, mining, and the tapping of water resources (Pohle and Gerique 2010). The concern about habitat loss in the region was recognized by Ecuadorian authorities in the creation of the Podocarpus National Park in 1982 and the later establishment of protective forest areas. Subsequently, in September 2007 UNESCO declared the region as the Biosphere Reserve Podocarpus El Cóndor.

The area is inhabited by different ethnic groups. The Jívaroan or Shuar indigenous groups are the traditional inhabitants of the humid premontane forests and lowlands of south-eastern Ecuador and north-west Peru. The Shuar represent one of the most important indigenous cultures in the Amazon Basin (Kingman 2005, Descola 1994). The Shuar communities try to maintain their own identity and language and most of them are still forest dwellers practicing shifting cultivation, mainly in a subsistence economy. During the last decades the colonization processes and the infrastructural opening of their territories has changed their way of life and their livelihoods (Pohle and Gerique 2006).

The Saraguros are Quechua Indians; their traditional territory spans the northern Andean highlands of the Province of Loja. Most of them are engaged in agro-pastoral activities that combine cattle ranching and horticulture and crop production (Pohle and Gerique 2010). The economic success brought by cattle breeding and an increase in their population gave rise to a shortage of pasture land in their home areas (Belote 1998). Consequently, the Saraguros are colonizing new territories toward the eastern regions to increase their pastures.

The mestizos are people of mixed Spanish and indigenous descent. In the study area they are the principal ethnic group in numbers; the majority arrived during the last decades from the western areas of Loja Province to practice agriculture and to log timber. The colonization process got decisive support from a severe drought in 1968, which affected one third of the population and resulted in the migration of thousands of ruined mestizo small farmers (Temme 1972). Figure 1 shows the areas of settlement of the different ethnic groups around the Podocarpus National Park and the investigated settlement areas.

## 3. Aims and Methods of this Study

Based on the foregoing, this study pursued the following objectives:

- To document the use of plant diversity by the different ethnic groups to avoid its loss through acculturation and other modernization processes and to identify potential novel plant species for food, fibers and other commodities.
- To describe and analyze current land use in order to identify weaknesses and strengths.
- To identify land use alternatives that fit in the area (focusing on improved agroforestry, ecotourism, PES, bioprospecting) to formulate recommendations following the concept of "protection by use."

We carried out the research project within the Research Units 402 (Functionality in a tropical Mountain Rainforest: Diversity, Dynamic Processes and Utilization Potentials under Ecosystem Perspectives) and 816 (Biodiversity and Sustainable Management of a Megadiverse Mountain Ecosystem in South Ecuador) of the German Research Foundation (DFG). We undertook fieldwork in sample communities of indigenous Shuar in the Upper Nangaritza as well as indigenous Saraguros and local mestizos in the Upper Zamora between March 2004 and November 2010, totaling nine visits to Ecuador and around 19 months of field research. As part of the DFG research units, the study fulfilled all legal research requirements of the Ecuadorian national authorities. Permission to work in Shuar territory was granted by the Shuar local authorities. The Saraguros and the mestizos were not represented by a central authority, thus individual permission for each investigated household was required.

During the ethnobotanical survey, we conducted 94 interview among 67 male and female informants. We used various techniques of unstructured and semi-structured inquiry, such as participant observation techniques, inventory technique, artifact and plant interviews, checklist interviews, and group interviews (cf. Alexiades 1996). We conducted all interviews in Spanish. In some cases it was not possible to obtain permission from local people to collect plant specimens. In order to provide a permanent record of information, pictures of almost all plant species were taken. We identified plant specimens with the help of Ecuadorian botanists of the Herbarium Reinaldo Espinosa of the Universidad Nacional (LOJA) and deposited them there.

To quantify ethnobotanical data, we chose the "use totalled method" (cf. Phillips 1996). The number of uses was totaled by catego-

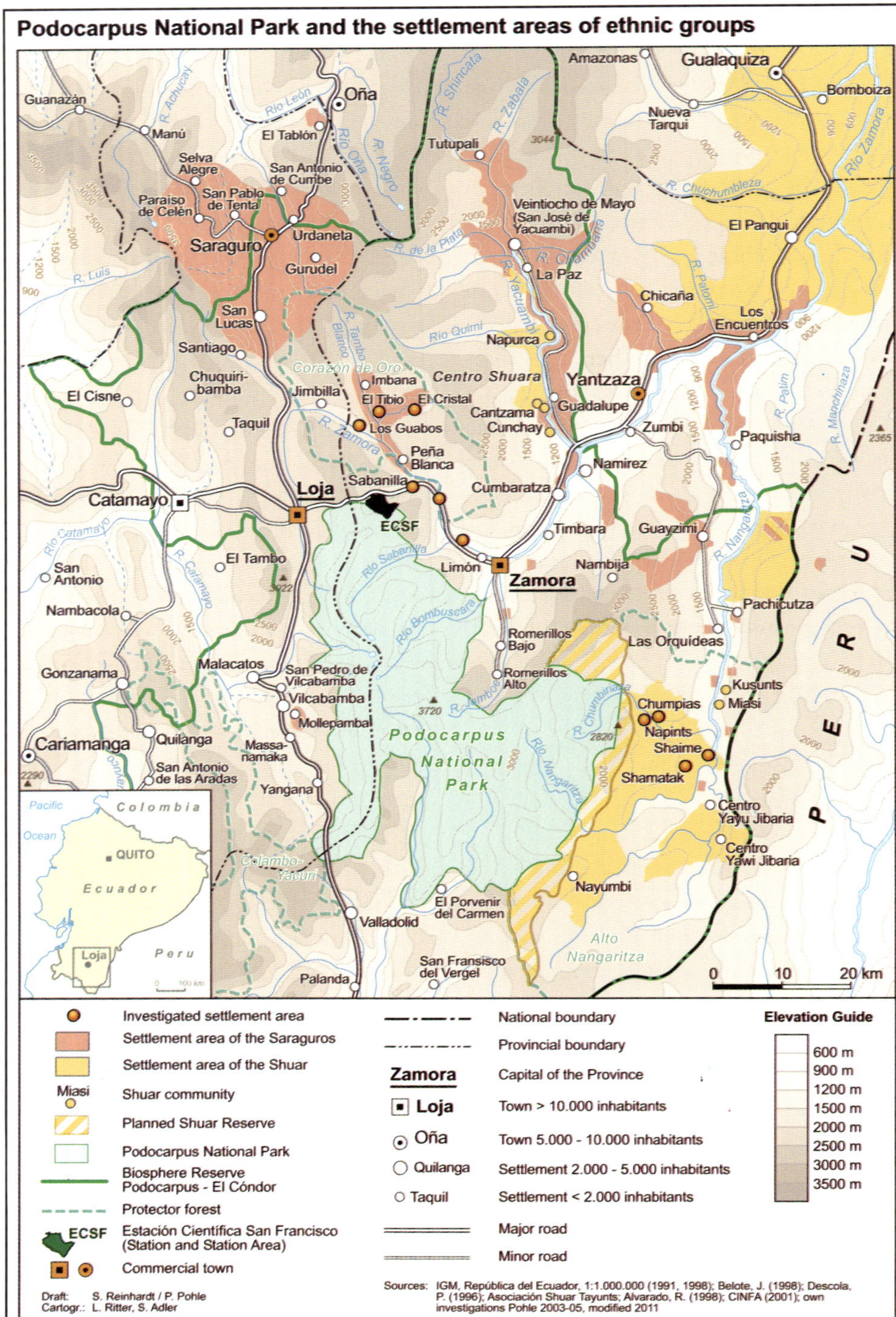


Figure 1. Study sites and areas of settlement of the different ethnic groups around the Podocarpus National Park, Ecuador. (Excerpted from Gerique 2010.)

ry of plant use, without evaluating the relative importance of each use. We established the use categories according to the main uses recorded and to the number of useful plant species with the same use among one ethnic group. At least five different plant species with the same use within the same ethnic group had to be recorded to establish a new use category. The categories are non-exclusive; one species can be represented in more than one use category if it has more than one use.

In order to highlight the significance of the forest products on the one hand and the biodiversity of the home gardens on the other, we divided plant species into two main categories: "wild plant species" and "cultivated plant species." In practice, the distinction was complicated to make, as there is a wide spectrum between completely wild and completely domesticated species. In such cases, the category indicated by the informants was accepted. Hence, the categories are not exclusive. In order to highlight the importance of forests as the place where the ethnic groups get plant resources, we divided wild species into two subjective sub-categories, namely wild species collected in secondary and mature forests and wild species harvested predominantly in fields, gardens, pastures and other disturbed areas. On the one hand, we considered forest species protected during forest clearing "wild species gathered in forest," even if the forest had disappeared. On the other hand, we considered native pioneer species that develop in disturbed environments after clearing "wild species gathered in other areas."

We collected information about the historical development, settlement patterns, local land use, and alternatives through semi-structured interviews and field observations. A literature survey on the use of agroforestry, ecotourism, payments for environmental services, and bioprospecting in Ecuador was followed by 15 semi-structured expert interviews conducted with professionals from different fields of activity in order to discuss the feasibility of such alternative land use and conservation options.

## 4. RESULTS

### 4.1 The ethnobotanical inventory

The ethnobotanical inventory encompasses 748 species, representing 132 families and 440 genera. The largest botanical families are Asteraceae (37), Solanaceae (36), and Poaceae (30). The botanical inventory includes 644 useful species with at least one use. Among the useful plant species, large cosmopolitan families (Asteraceae, Solanaceae, Poaceae, Fabaceae, Lamiaceae) and large tropical families (Euphorbiaceae, Araceae, Piperaceae, Lauraceae, Melastomataceae) are the most frequent in south-eastern Ecuador.

#### *4.1.1 Results of the ethnobotanical survey in Shuar communities*

Within the inventory, 316 different plant species with a total of 493 uses were recorded. Among the species, 211 are wild plants and 113 are cultivated plants, with eight species being represented in both categories. Of the recorded useful specimens, 305 (97%) were identified to genus, and 188 (60%) to species or species affinity. These are distributed among 84 families and 211 genera. The family, which includes the most useful species, is Solanaceae with 19 species, followed by Araceae with 17, Piperaceae with 15 and Poaceae and Fabaceae with 14 species each. The Family Importance

Value Index (FIVI) by Báez and Borgtoft (FIVI = $Usp_1 + Usp_2 + \ldots + Usp_n$, cf. Báez 1999) rates Arecaceae (FIVI = 32), Solanaceae (FIVI = 28), Araceae (FIVI = 22), Fabaceae (FIVI = 20), and Piperaceae (FIVI = 20) as the most important ethnobotanical families among the Shuar. Other ethnobotanical studies of the Shuar of the Nangaritza Valley (cf. Byg 2004, Santín 2004, Van den Eyden 2004) also identify the importance of the Arecaceae. Palms are one of the most important forest resources for indigenous peoples throughout the Amazon region (Byg and Balslev 2006). The investigation underlines the Shuar's vast knowledge of plants and their uses.

As shown in Figure 2, sixteen use categories appear. The most relevant category is "medicinal plants," with 105 species. The second most important is "food" with 100 plant species. Plants cultivated in forest gardens provide most of the calorific intake of the Shuar, while other species cultivated in gardens and collected in the forest allow for dietary variety. Several species (67), most of them wild species (43), provide the Shuar with materials for the construction of houses, furniture, and canoes. The high number in the category fodder (44) results from wild plant species eaten by game, as the Shuar consider such plants to be fodder (Pohle et al. 2010, Bennett et al. 2002). This category is followed by plants used for fuel (30), ornamental (22), tools and baskets (22), hunting and fishing (17), crafts (15), ritual and mythical (15), veterinary (15), paint, dye and varnish (9), fibers (7), and breeding of edible palm weevils (5). The latter is a com-

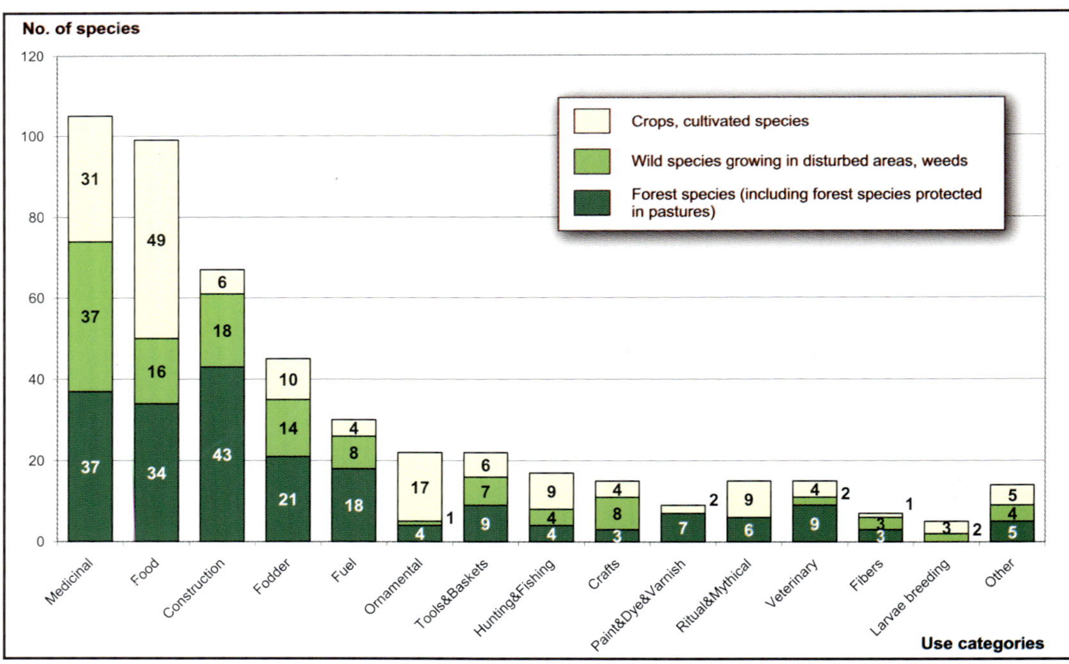


Figure 2. Shuar plant use categories with at least five species. The bars have been divided according to their origin (cultivated or wild) and, in the case of wild species, in accordance with their main place of harvesting. (Excerpted from Gerique 2010.)

Note: The categories are non-exclusive; species with more than one use are included in more than one category.

mon practice among the indigenous peoples of the Amazon and provides an important dietary supplement (Choo et al. 2009). Further 14 species with other uses (such as shampoos, perfumes, poisons, fertilizers, living fences, or insecticides) have been listed under the category "other uses."

### *4.1.2 The ethnobotany of the Saraguros*

The ethnobotanical inventory of the Saraguros comprises 230 plant species with a total of 310 uses. Of these, 125 plants are wild and 113 are cultivated species, while eight species are represented in both categories. Of the recorded specimens, 99% have been identified to genus and 80% to species or species affinity. These belong to 88 families and to 181 genera. Asteraceae is the most represented family with 19 species, followed by Poaceae and Solanaceae with 17 species. The Rosaceae are represented with 10 species. Using the Family Importance Value Index (FIVI) after Báez and Borgtoft (1999), the most important families are Solanaceae and Asteraceae, (both FIVI = 22) and Poaceae (FIVI = 20).

Nine use categories appear after grouping the plant uses. Figure 3 underlines the importance of plant species for the Saraguros as food (84 species) and medicine (75) sources. The subsequent categories include species used as ornaments and decoration (39), in construction (37), fodder (22), fuel (13), living fences (13), tools and baskets (10), or for

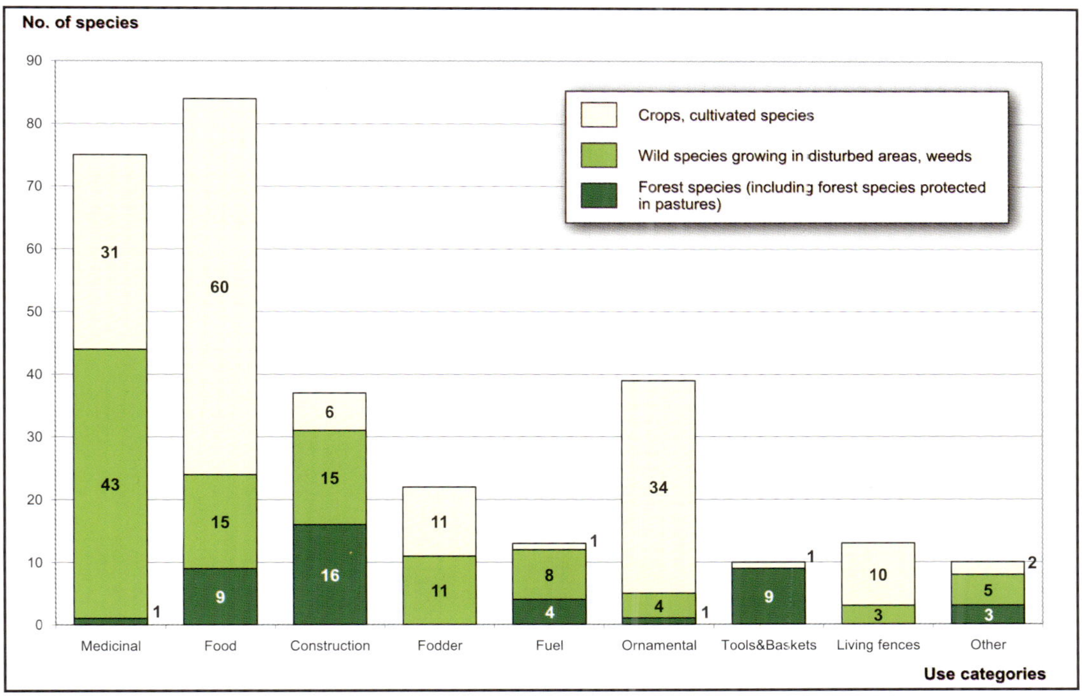


Figure 3. Saraguro use categories with at least five species. The bars have been divided according to their origin (cultivated or wild) and, in the case of wild species, in accordance with their main place of harvesting. (Excerpted from Gerique 2010.)
Note: The categories are non-exclusive; species with more than one use are included in more than one category.

"other" uses (10). The latter category includes plants used to make cords, and uses such as veterinary, dyes, or religious ceremonies.

### *4.1.3. Results of the ethnobotanical survey in mestizo communities*

The mestizos who live in the research know 312 species with a total of 409 uses. Of them, 148 are wild plants and 172 are cultivated plants. Eight species were included in both categories. A total of 98% of species (306) have been identified to genus and 78% (242) to species or species affinity. The useful species belong to 101 botanical families and at least to 237 genera. Asteraceae is the most represented family (22 useful species), followed by Poaceae (19), Solanaceae (13), and Rosaceae (12). According to the Family Importance Value Index (FIVI) after Báez and Borgtoft (Báez 1999) the most important families are Asteraceae (with FIVI = 28), Poaceae (with FIVI = 22), and Solanaceae (FIVI = 18).

As in the case of the Saraguros, the mestizo ethnobotany consists of nine use categories, which are represented in Figure 4. The most important use categories are "food" with 103 plant species, "medicinal," with 99 plant species used to treat different ailments, and ornamental (78). Like the Saraguros, the mestizos make a profuse use of plants to decorate their houses, home gardens, and to embellish the local chapels. Other categories are construction (37 species), fodder (29), living fences

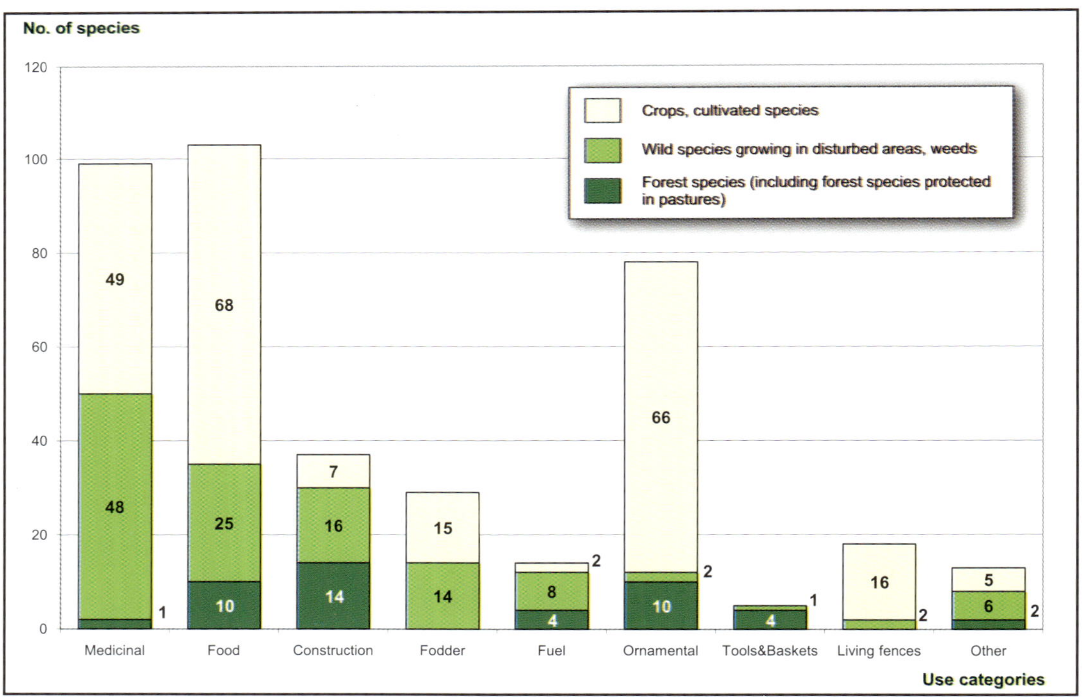


Figure 4. Mestizo use categories with at least five species. The bars have been divided according to their origin (cultivated or wild) and, in the case of wild species, in accordance with their main place of harvesting. (Excerpted from Gerique 2010.)

Note: The categories are non-exclusive; species with more than one use are included in more than one category.

(18), fuel (14), tools and baskets (5). The category "other" encompasses 13 plant species used for veterinary and religious/ritual purposes, and dyes.

### *4.1.4 The relevance of forests as areas of harvesting among the ethnic groups*

A look at the sites where plants are principally collected shows the importance of the resource "forest" among the different ethnic groups. Figure 5 represents the differences. The yellow section of the bars shows the percentage of cultivated species in fields and home gardens. The percentage of useful wild species that are commonly collected in fields, gardens, pastures and other disturbed areas appears in light green, while the percentage of useful species usually gathered in secondary and mature forests, and of forest species that grow protected in pastures after forest clearing, is represented in dark green.

One-third (34%) of the plants used by the Shuar are cultivated plants in home and forest gardens. For the most part, they serve as food and medicines. Almost 42% of the plant species are collected in the forest, which provides a "safety net" for the Shuar's livelihoods, attenuating the lack of direct access to markets and to a formal healthcare system (Pohle and Reinhardt 2004). Most construction materials are collected in forests (31%). Also, the forest

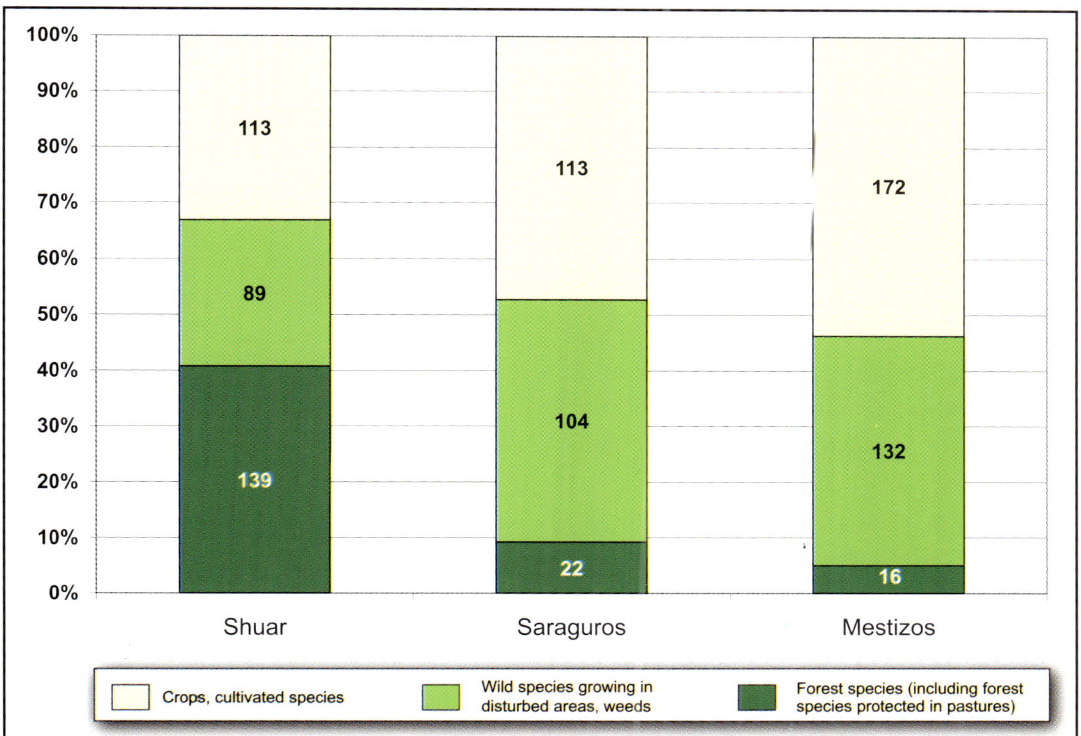


Figure 5. Plant species used by Shuar, Saraguros, and Mestizos according to their gathering places. (Excerpted from Gerique 2010.)
Note: The numbers inside the columns indicate total values and not percentages.

offers the Shuar easy access to wood used for fuel, to most of the species known as fodder for game, and to species needed to make dyes and paints and for veterinary use. Furthermore, around 24% of the useful species are collected in disturbed sites. These areas are important for the supply of woody species used in construction, and for medicinal plants.

The Saraguros and the mestizos get around 50% of the plants (including fodder species in pastures) in their fields and home gardens. Food, ornamental plants, and medicinal plants are the main needs covered met through cultivation. In contrast, the Saraguros and the mestizos make very limited use of forest species. Only 9% of the species used by the Saraguros and 5% of the species used by the mestizos are forest plants. Most of them are timber species. The remaining forest species either do not play an important role in their livelihoods (culturally or economically), or can be easily substituted by other "non-forest species" that grow closer to the households, or by manufactured products. On the other hand, disturbed areas and pastures are quite important as collection sites. Around 42% of the useful plants are gathered here. As in the case of the Shuar, numerous medicinal plants are collected in such areas. The Saraguros and mestizos collect around 97% of the wild growing medicinal plants in disturbed sites and pastures. The results support the position of Stepp and Moerman (2001), who argued that the trend of searching for new drugs in primary tropical forest areas suggests that other habitats, such as disturbed areas, have been ignored by science so far. Furthermore, these ethnic differences in the relevance of forests should be taken into consideration when developing alternative land uses.

## 4.2 Land Use in South-eastern Ecuador

### *4.2.1 Land Use in the Shuar Settlements*

Similar to other areas inhabited by the Shuar (cf. Duchelle 2007, Rudel et al. 2002, Borgtoft et al. 1999), land use in the Upper Nangaritza results in a complex and heterogeneous array of spaces that change with decreasing intensity of use. Home gardens are located adjacent to or surrounding households; close to and around the settlements the Shuar land use creates a mosaic of new and old forest gardens and pastures inside secondary forest areas in different stages of the regeneration process, and patches of mature forest with limited impact from human activities. A few hours away from the settlements human pressure declines and the landscape appears dominated by mature forest.

The home gardens and forest gardens are places of great agrobiodiversity and refuges of genetic resources (Coomes and Ban 2004, Kumar and Nair 2004). They have a low impact on forest vegetation due to the small size of the plots, while the presence of wild species facilitates a rapid regeneration of vegetation cover after use. In the Upper Nangaritza, the fallow periods last for more than 20 years, while the cultivation periods cover four or five years. Likewise, the fertilization of soils by mulching and natural fertilizers, the use of traditional tools like digging sticks, and the utilization of natural insecticides makes it valid to include the Shuar among smallholders who practice ecologically sustainable agriculture (Pohle and Gerique 2006, Rudel et al. 2002).

However, the Shuar are definitively not "ecologically noble savages" (cf. Alvard 1993). Their increasing dependence on market economy has led to new, non-traditional production systems. They raise cattle and have established and expanded their pastures. The possible over-use of certain non-timber forest products such as palm hearts remains unclear and they have taken part in the over-exploitation of logging. Moreover, in recent times they have intensified crop production; some Shuar use now large plots and agrochemicals to produce *Solanum quitoense* Lam., making the future of traditional shifting cultivation methods uncertain. The Shuar have apparently over-exploited game species as well, a factor that may have consequences in plant diversity as it could alter zoochory negatively. Yet, most of the activities that lead to significant changes in land cover and jeopardize local biodiversity in the short-time have been imported from outside their communities. Deforestation by settlers, the construction of roads, and gold mining are the main threats to biodiversity conservation in the Upper Nangaritza. Especially the latter two appear as imminent threats. A new road that crosses the valley is being constructed. Furthermore, prospecting rights are under process for about 80% of the Upper Nangaritza (UNL et al. 2006). A first serious attempt at large-scale prospecting failed in 2007 due to the armed resistance of the Shuar. However, many Shuar have begun to pan for gold.

#### *4.2.2 Land Use among the Saraguros and Mestizos*

The Saraguros and the mestizos of the Upper Zamora share plant and land use. Both ethnic groups have transformed most of the pristine vegetation of their communities into home gardens, crop fields, pastures, and forest fragments, the latter located mostly in the upper part of the hills or in ravines. Fallows and wastelands in degraded or steep areas—often dominated by the bracken fern (*Pteridium arachnoideum* (Kaulf.) Maxon)—complete the landscape (cf. Pohle et al. 2010, Pohle and Gerique 2006).

Subsistence is guaranteed by cultivation of maize and beans in fields and the production of vegetables and fruits in home gardens; as in the case of the Shuar, home gardens host a high agrobiodiversity. However, there is an extreme variability from one settler to the next in their way of managing home gardens and land, especially among mestizo settlers. Livestock production is market oriented and represents the main land use system; it fulfills multiple objectives: the production of beef and dairy products provides households with a regular income; it awards a prestigious social status, and represents a way of accumulating wealth.

As pointed out already, with the exception of timber the Saraguros and mestizos make little use of forest resources. As pointed out by Wunder (1996), what matters is the potential for agricultural use of the underlying soil. Ranchers set fires to open new pastures and to rejuvenate them. Uncontrolled fires due to negligence or the burning of more land than the farmer can hold under production leads to substantial loss of forest and to the growth of large wastelands. Moreover, the construction of roads improves transport for livestock production, making farming more profitable and leading to the clearing and planting of larger areas. The described land use results not only in the substitution of forest with pastures but also in a high fragmentation of the remaining forest areas and their disturbance with subsequent edge effects. In this way the Saraguros and mestizos jeopardize the sustainability of their economic systems.

## 5. Alternative Land Use Practices for Biodiversity Conservation

### 5.1 Alternatives for the Upper Nangaritza

The Shuar depend on forest resources are closely bound culturally, spiritually, and economically to the forest (Pohle and Gerique 2006). Therefore, they are interested in forest conservation. The challenge is to avoid the Shuar transforming their well-adapted traditional land use towards more intensive production systems that could endanger local forests. As a first step non-traditional, non-sustainable practices should be removed and prevented, and additional sustainable sources of income should be introduced.

Agroforestry systems together with community-based ecotourism appear to be the most promising sustainable sources of alternative income. From a local commercial perspective, *Terminalia amazonia* (J.F. Gmel.) Exell, *Platymiscium pinnatum* (Jacq.) Dugand, and *Cedrelinga cateniformis* (Ducke) Ducke are interesting species for reforestation, while the afforestation of pastures should include *Trema micrantha* Blume, Inga spp. (Revelo and Palacios 2005, Vázquez-Yanes 1998), *Oenocarpus bataua* Mart., *Bactris gasipaes* Kunth (Miller 2002, Van den Eyden et al. 1998), *Cordia alliodora* (Ruiz & Pav.) Oken, and *Cedrela odorata* L. (Camargo 2003, Somarriba et al. 2001). The production of organic coffee (*Coffea arabica* L.), naranjillas (*Solanum quitoense* Lam.), and cocoa (*Theobroma cacao* L.), should be combined with reforestation and afforestation. Many Shuar have experience with these species, and demand is high. The German Technical Assistance GIZ has recently showed interest in the production of such crops in the Upper Nangaritza.

Ecotourism has a great potential in the area, as it can be considered one of the most beautiful landscapes in the country (Palacios 1997). Own interviews revealed that local attractions such as the presence of indigenous groups, flora, and archeological sites are among the most demanded attractions by tourists who visit south-eastern Ecuador. Finally, other options such as payments for environmental services and bioprospecting ventures appear less attractive in the short and medium scale, mainly because they are hostage to political interests, bad image and unclear legislation.

Due to the lack of income opportunities, the poverty in Ecuador, and the high price of gold in world markets, gold mining will remain one of main threats to biodiversity in the Upper Nangaritza. The role of the Ecuadorian Government and the future impact of international mining companies in the region go beyond the scope of this study. Nevertheless, NGOs and authorities should take into consideration to help the Shuar to deal with the consequences of mining. There exist positive examples of projects that have minimized the impacts artisanal gold-mining activity (Mesa et al. 2010, Sousa and Veiga 2009) that could be developed in the area.

### 5.2 Alternatives for the Upper Zamora

In the case of the Saraguros and mestizos, the actual land use system based on cattle ranching turns out to be incompatible with the conservation of forest resources and long term household prosperity (Pohle et al. 2010). Its significance in local economies should thus be reduced through diversification.

An agroforestry system that includes reforestation, restoration and connection of forest patches, and sustainable selective logging could be implemented in the medium term. Plantations of *Pinus patula* Schiede & Deppe ex Schltdl. can restore forest cover and produce revenues within acceptable times (Brockerhof et al. 2008). In subsequent rotations, native species (such as *Alnusa cuminata* Kunth, *Cedrela* spp., and *Tabebuia chrysantha* (Jacq.) G. Nicholson) could be introduced to these plots, as *Pinus patula* plantations also have a nurse effect for native species, facilitating the reestablishment of more natural forest sites (Weber et al. 2008, Aguirre et al. 2006). Following the recommendations by Lamb et al. (2005) degraded pastures close to forest remnants could be reforested with a mixture of commercially attractive native and native species that otherwise would be unable to colonize and regenerate or are ecologically vulnerable. In the short term, improved pasture management, the production of honey (cf. Ordoñez and Lalama 2006), and the small-scale production of niche products (such as ornamental plants) in home gardens could be alternatives to cattle ranching (Pohle et al. 2010).

Payments for environmental services (e.g., payments for watershed protection or the Ecuadorian Socio Bosque Program) are an example of a way of generating the seed money for such projects. If farmers could get compensation payments, the required investment to establish and maintain an agroforestry system would shrink considerably (Knoke et al. 2009), and the abundance of low-productive lands and the large options for the intensification of production make leakage (leakage refers to deforestation processes which occur outside the agroforestry system but are ascribable to its activities) improbable in the study area (cf. Wunder and Albán 2008). Finally, ecotourism could be another tool to improve local income. Even though the income may be small in comparison to profit in other areas, the benefits would represent a considerable rise in local purchasing power.

## LITERATURE CITED


Aguirre, N., S. Günter, M. Weber, and B. Stimm. 2006. Enrichment of *Pinus patula* Plantations with Native Species in Southern Ecuador. *Lyonia* 10(1):33–45.

Alexiades, M.N. (ed). 1996. *Selected Guidelines for Ethnobotanical Research: A Field Manual.* The New York Botanical Garden Press, Bronx, NY.

Alvard, M.S. 1993. Testing the "Ecologically Noble Savage" Hypothesis: Interspecific Prey Choice by Piro Hunters of Amazonian Peru. *Hum Ecol* 21(4):355–387.

Báez, S. 1999. El uso de los recursos del bosque. Pp. 121–132 in H. Borgtoft, F. Skov, J. Fjedså, I. Schjellerup, and B. Øllgaard (eds.), *La Gente y la Biodiversidad. Dos Estudios en Comunidades de las Estribaciones de los Andes en Ecuador.* Centro para la Investigación de la Diversidad Cultural y Biológica de los Bosques Pluviales Andinos, Dinamarca y Ediciones Abya Yala, Quito.

Barthlott, W., A. Hostert, G. Kier, W. Küper, H. Kreft, J. Mutke, D. Rafiqpoor, and H. Sommer. 2007. Geographic Patterns of Vascular Plant Diversity at Continental to Global Scales. *Erdkunde* 61(4):305–315.

Belote, J. 1998. *Los Saraguros del Sur del Ecuador.* Abya Yala. Quito.

Bhagwat, S.A., J. Willis, H.J. Birks, and R.J. Whittaker. 2008. Agroforestry: a Refuge for Tropical Biodiversity? *Trends Ecol Evol* 23(5):261–267.

Bennett, B.C., M.A. Baker, and P. Gómez Andrade. 2002. *Ethnobotany of the Shuar of Eastern Ecuador.* Advances in Economic Botany Volume 14. The New York Botanical Garden Press, Bronx, NY.

Borgtoft, H., F. Skov, J. Fjedså, I. Schjellerup, and B. Øllgaard (eds.). 1999. *La Gente y la Biodiversidad. Dos Estudios en Comunidades de las Estribaciones de los Andes en Ecuador.* Centro para la Investigación de la Diversidad Cultural y Biológica de los Bosques Pluviales Andinos, Dinamarca y Ediciones Abya Yala, Quito.

Brehm, G., J. Homeier, K. Fielder, I. Kottke, J. Illig, N.M. Nöske, F. Werner, and S.-W. Breck. 2008. Mountain Rain Forests in Southern Ecuador as a Hotspot of Biodiversity—Limited Knowledge and Diverging Patterns. Pp. 15–24 in E. Beck, J. Bendix, I. Kottke, F. Makeschin, and R. Mosandl (eds.), *Gradients in a Tropical Mountain Ecoystem of Ecuador.* Springer-Verlag, Berlin Heidelberg.

Brockerhoff, E.G., H. Jactel, J.A. Parrotta, C.P. Quine, and J. Sayer. 2008. Plantation Forests and Biodiversity: Oxymoron or Opportunity. *Biodivers Conserv* 17:925–951.

Bruner, A.G., R.E. Gullison, R.E. Rice, and G.A.B. da Fonseca. 2001. Effectiveness of Parks in Protecting Tropical Biodiversity. *Science* 291:125–128.

Byg, A. 2004. *Humans and Plants of the Rain Forest.* Ph.D. Thesis. Department of Systematic Botany, University of Aarhus.

Byg, A., and H. Balslev. 2006. Palms in Indigenous and Settler Communities in Southeastern Ecuador: Farmers' Perceptions and Cultivation Practices. *Agroforest Syst* 67:147–158.

Camargo, J.C. 2003. Sistemassilvopastoriles con Laurel (*Cordia alliodora*) en el trópico húmedo y sub-húmedo de Costa Rica. In *II Conferencia Electrónica sobre Agroforestería para la Producción Animal en América Latina.* FAO, accessed September 30 2011, http://www.fao.org/ag/aga/AGAP/FRG/agrofor1/Agrofor1.htm.

Chazdon, R.L. 1998. Tropical Forests—Log 'Em or Leave 'Em? *Science* 281(5381):1295–1296.

Choo, J., E.L. Zent, and B.B. Simpson. 2009. The Importance of Traditional Ecological Knowledge for Palm-Weevil Cultivation in the Venezuelan Amazon. *J. Ethnobiol* 29(1):113–128.

Cincotta, R.P., J. Wisnewski, and R. Engelman. 2000. Human Population in the Biodiversity Hotspots. *Nature* 404:990–991.

Colchester, M. 1996. Beyond "Participation": Indigenous Peoples, Biological Diversity Conservation and Protected Area Management. *Unasylva - Number 186: Forest Dependant People.* Vol. 47(3). FAO, accessed September 30 2011, http://www.fao.org/docrep/w1033e/w1033e00.htm

———. 2000. Self-Determination or Environmental Determinism for Indigenous Peoples in Tropical Forest Conservation. *Conserv Biol* 14(5):1365–1367.

Coomes, O.T., and N. Ban. 2004. Cultivated Plant Species Diversity in Home Gardens of an Amazonian Peasant Village in Northeastern Peru. *Econ Bot* 58(3):420–434.

Daily, G.C., and K. Ellison. 2002. *The New Economy of Nature. The Quest to Make Conservation Profitable.* Island Press, Washington, DC.

Descola, P. 1994. *In the Society of Nature: A Native Ecology in Amazonia.* Cambridge University Press, Cambridge, UK.

Duchelle, A.E. 2007. Observations on Natural Resource use and Conservation by the Shuar in Ecuador's Cordillera del Cóndor. *Ethnobot Res & Appl* 5:5–23.

Durand, L., and E. Lazos. 2008. The Local Perception of Tropical Deforestation and its Relation to Conservation Policies in Los Tuxtlas Biosphere Reserve, Mexico. *Hum Ecol* 36:383–394.

Forest Trends, The Katoomba Group, and UNEP. 2008. *Payments for Ecosystem Services. Getting Started: A Primer.* UNON/Publishing Services Section. Nairobi.

Gerique, A. 2010. *Biodiversity as a Resource: Plant Use and Land Use among the Shuar, Saraguros, and Mestizos in Tropical Rainforest Areas of Southern Ecuador.* Dissertation. Institute of Geography, University of Erlangen-Nuremberg.

Gray, C.L., R.E. Bilsborrow, J.L. Bremmer, and F. Lu. 2008. Indigenous Land Use in the Ecuadorian Amazon: A Cross-Cultural and Multilevel Analysis. *Hum Ecol* 36:97–109.

Günter, S., P. González, G. Álvarez, N. Aguirre, X. Palomeque, F. Haubrich, and M. Weber. 2009. Determinants for Successful Reforestation of Abandoned Pastures in the Andes: Soil Conditions and Vegetation Cover. *Forest Ecol Manag* 258:81–91.

Kingman, S. 2005. Historias Equidistantes. Deriva de Colonos y Shuar. *Ecuador Terra Incognita* 36:22–36.

Knoke, T., M. Weber, J. Barkmann, P. Pohle, B. Calvas, C. Medina, N. Aguirre, S. Günter, B. Stimm, R. Mosandl, F. von Walter, B. Maza, and A. Gerique. 2009. Effectiveness and Distributional Impacts of Payments for Reduced Carbon Emissions from Deforestation. *Erdkunde* 63(4):365–384.

Kumar, B.M., and P.K.R. Nair. 2004. The Enigma of Tropical Home Gardens. *Agroforest Syst* 1:135–152.

Kursar, T.A., C.C. Caballero-George, T.L. Capson, L. Cubilla-Ríos, W.H.G. Gerwick, M.V. Heller, A. Ibañez, R.G. Linington, K.L. McPhail, E. Ortega-Barría, L.I. Romero, and P.D. Coley. 2007. Linking Bioprospecting with Sustainable Development and Conservation: The Panama Case. *Biodivers Conserv* 16:2789–2800.

Lamb, D., P.D. Erskine, and J.A. Parrotta. 2005. Restoration of Degraded Tropical Forest Landscapes. *Science* 310:1628–1632.

Mesa, C., P. Alfonso, E. Monterde, and M. Costa. 2010. Cooperación en el Campo de la Pequeña Minería en Sudamérica: El Papel de las ONGs. *Boletín de la Sociedad Geológica Mexicana* 62(1):109–122.

Miller, C. 2002. Fruit Production of the Ungurahua Palm (*Oenocarpusbataua* subsp. *bataua*, Arecaceae) in an Indigenous Managed Reserve. *Econ Bot* 56(2):165–176.

Myers, N., R. Mittermeier, C. Mittermeier, G.A.B. da Fonseca, and J. Kent. 2000. Biodiversity Hotspots for Conservation Priorities. *Nature* 403:853–858.

Ordoñez, O., and K. Lalama. 2006. *Experiencias del Manejo Apícola en Uritusinga. Loja - Ecuador.* Fundación Ecológica Arco Iris, PROBONA y Asociación de Productores Agropecuarios Arco Iris. Loja, Ecuador.

Palacios, W.A. 1997. Botany and Landscape of the Río Nangaritza Basin. Pp. 37–44 in T.S. Schulenberg and K. Awrey (eds.), *The Cordillera del Cóndor Region of Ecuador and Peru: A Biological Assessment.* RAP Working Papers, No. 7. Conservation International, Washington, DC.

Phillips, O. 1996. Some Quantitative Methods for Analysing Ethnobotanical Knowledge. Pp. 171–197 in M.N. Alexiades (ed.), *Selected Guidelines for Ethnobotanical Research: A Field Manual.* The New York Botanical Garden, Bronx, NY.

Pohle, P., and A. Gerique. 2006. Traditional Ecological Knowledge and Biodiversity Management in the Andes of Southern Ecuador. *Geographica Helvetica, Swiss Journal of Geography* 4:275–285.

———. 2008. Use of Biodiversity along the Altitudinal Gradient: Plant Knowledge and Plant Use among Indigenous Shuar and Saraguro Communities in Southern Ecuador. Pp. 331–346 in E. Beck, J. Bendix, I. Kottke, F. Makeschin, and R. Mosandl (eds.), *Gradients in a Tropical Mountain Ecoystem of Ecuador.* Springer-Verlag, Berlin Heidelberg.

Pohle, P., and S. Reinhardt. 2004. Indigenous Knowledge of Plants and Their Utilization among the Shuar of the Lower Tropical Mountain Forest in Southern Ecuador. *Lyonia* 7(2):33–149.

Pohle, P., A. Gerique, M. Park, and M.F. López Sandoval. 2010. Human Ecological Dimensions in Sustainable Utilization and Conservation of Tropical Mountain Rain Forests under Global Change in Southern Ecuador. Pp. 477–509 in T. Tscharntke, C. Leuschner, E. Veldkamp, H. Faust, E. Guhardja, and A. Bidin (eds.), *Tropical Rainforests and Agroforests under Global Change.* Springer-Verlag, Berlin Heidelberg.

Revelo, N., and W.A. Palacios. 2005. *Avances Silviculturales en la Amazonía Ecuatoriana: Ensayos en la Estación Biológica Jatun Sacha.* Fundación Jatun Sacha & Proyecto CAIMAN. Quito.

Robertson, N., and S. Wunder. 2005. *Fresh Tracks in the Forest: Assessing Incipient Payments for Environmental Services Initiatives in Bolivia.* CIFOR, Bogor.

Rudel, T.K., D. Bates, and R. Machinguiashi. 2002.Ecologically Noble Amerindians? Cattle Ranching and Cash Cropping among Shuar and Colonists in Ecuador. *Lat Am Res Rev* (37)1:144–159.

Santín, F. 2004. Ethnobotany of the Communities of the Upper Río Nangaritza. *Lyonia* 7(2):105–122.

Sayer, J., N. Ishwaran, J. Thorsell, and T. Sigaty. 2000. Tropical Forest Biodiversity and the World Heritage Convention. *Ambio* 29(6):302–309.

Sinclair, A.R.E., D. Ludwig, and C.W. Clark. 2000. Conservation in the Real World. *Science* 289:1875–1876.

Somarriba, E., R. Valdivieso, W. Vásquez, and G. Galloway. 2001. Survival, Growth, Timber Productivity and Site Index of *Cordia alliodora* in Forestry and Agroforestry Systems. *Agroforest Syst* 51:111–118.

Sousa, R.N., and M.M. Veiga. 2009. Using Performance Indicators to Evaluate an Environmental

Education Program in Artisanal Gold Mining Communities in the Brazilian Amazon. *Ambio* 38(1):40–46.

Stepp, J.R., and D.E. Moerman. 2001. The Importance of Weeds in Ethnopharmacology. *J Ethnopharmacol* 75:19–23.

Temme, M. 1972. *Wirtschaft und Bevölkerung in Südecuador. Eine Sozio-Ökonomische Analyse des Wirtschaftsraum Loja.* Kölner Forschungen zur Wirtschafts- und Sozialgeographie (Band 19). F. Steiner, Wiesbaden.

Terborgh, J., and C.A. Peres. 2002. The Problem of People in Parks. Pp. 307–319 in J. Terborgh, C. van Schaik, L. Davenport, and M. Rao (eds.), *Making Parks Work: Strategies for Preserving Tropical Nature.* Island Press, Washington, DC.

Universidad Nacional de Loja, 2006. Estado de Conservación del Área de Bosque y Vegetación Protectora "Cuenca Alta del Río Nangaritza." Pp. 104–132 in Universidad Nacional de Loja (ed.) *Estado de Conservación de Áreas Protegidas y Bosques Protectores de Loja y Zamora Chinchipe y Perspectivas de Intervención.* Loja, Ecuador.

Van den Eynden, V. 2004. *Use and Management of Edible Non-Crop Plants in Southern Ecuador.* Ph.D. Thesis in Applied Biological Sciences, Ghent University.

Van den Eyden, V., E. Cueva, and O. Cabrera. 1998. *Plantas Silvestres Comestibles del Sur del Ecuador – Wild Edible Plants of Southern Ecuador.* Ediciones Abya Yala. Quito.

Van Schaik, C., and H.D. Rijksen. 2002. Integrated Conservation and Development Projects: Problems and Potential. Pp. 15–29 in J. Terborgh, C. van Schaik, L. Davenport, and M. Rao (eds.), *Making Parks Work: Strategies for Preserving Tropical Nature.* Island Press, Washington, DC.

Vázquez-Yanes, C. 1998. *Trema micrantha* (L.) Blume (Ulmaceae): A Promising Neotropical-Tree for Site Amelioration of Deforested Land. *Agroforest Syst* 40:97–104.

Weber, M., S. Günter, N. Aguirre, B. Stimm, and R. Mosandl. 2008. Reforestation of Abandoned Pastures: Silvicultural Means to Accelerate Forest Recovery and Biodiversity. Pp. 447–457 in E. Beck, J. Bendix, I. Kottke, F. Makeschin, and R. Mosandl (eds.), *Gradients in a Tropical Mountain Ecoystem of Ecuador.* Springer-Verlag, Berlin Heidelberg.

Wittemyer, G., P. Elsen, W.T. Bean, A. Coleman, C.O. Burton, and J.S. Brashares. 2008. Accelerated Human Population Growth at Protected Area Edges. *Science* 231:123–126.

World Bank 2006. *World Development Report 2007: Development and the Next Generation.* World Bank, Washington, DC.

Wunder, S. 1996. Deforestation and the Uses of Wood in the Ecuadorian Andes. *Mt Res Dev* 16(4):367–382.

Wunder, S., and M. Albán. 2008. Decentralized Payments for Environmental Services: The Cases of Pimampiro and PROFAFOR in Ecuador. *Ecol Econ* 65:685–698.

# The Healing Forests of Richard Evans Schultes: Inspiration for Conservation and the Search for New Drugs Derived from Ethnobotanical Research on Tropical Forest Plants Used in Traditional Medicine.

Steven R. King[1]

The legacy of Richard Evans Schultes has so many facets it is difficult to do justice to the myriad of people, organizations, forests, and countries that he impacted. He had a profound influence on my life and the development of the soon to be approved new pharmaceutical drug *crofelemer* for the treatment of HIV/AIDS chronic diarrhea. Dr. Schultes was one of the inspirations for, and a founding scientist of, the Shaman Pharmaceuticals Scientific Strategy Team. He provided critical guidance for the work of Shaman Pharmaceuticals from 1989 until his passing in 2001. He was a vocal advocate for the approach of short, medium, and long term reciprocity conducted by Shaman with all of it collaborators. He also strongly supported the creation of the non-profit benefit sharing organization created by Shaman to return a portion of the profits of any drugs to the countries and cultures which collaborated with Shaman field research teams. The classic book, *The Healing Forest*, authored by Dr. Schultes and his friend and colleague Robert Raffauf, inspired the name for this non-profit organization, The Healing Forest Conservancy. The book documents close to 1500 species of medicinal plants utilized in the Northwest Amazon. Dr. Schultes focused on many aspects of the biocultural diversity of tropical forests and indigenous knowledge. He was also a very gifted photographer, and his photographs have been published in several books. The most notable is a wonderful book of his exquisite black and white photographs entitled *The Lost Amazon: The Photographic Journey of Richard Evans Schultes* (2004), which was produced by Wade Davis, whose chronicle of Dr. Schultes life, *One River* (1997), is mentioned in the contribution by Michael Balick in this volume.

My introduction to Richard Evans Schultes was similar to the experience of many other people. I returned to the United States after living with an indigenous community of Angotere Secoya people in the Northern Peruvian Amazon in 1979. I had lived among this community for nine months, studying their diet and medicinal plants as part of a contract with a Peruvian non-governmental organization. I visited the Field Museum of Natural History in Chicago, where I had the good fortune to meet Timothy Plowman, one of Schultes greatest students. Tim became a lifelong friend and mentor, and he also directed me to visit Dr. Schultes at the Botanical Museum on way back to college in Maine. When I visited Schultes and described where I had been conducting fieldwork, he immediately stated that I had been working in "his country with his people," as the Secoya are closely related to several cultural groups that he worked with in adjacent regions of Colombia. Dr. Schultes also became for me, as did for so many young scientists, a lifelong mentor and continuing source of inspiration.

I tried to become a graduate student of Dr. Schultes, but by the time I received my degree in Human Ecology, he was working part time and was no longer accepting graduate students. He suggested I pursue my graduate studies with one of his students, Micheal Balick, and another colleague of his, Ian Prance. I took his advice and continued my ethnobotanical

[1] Senior Vice President Sustainable Supply, Ethnobotanical Research and IP, Napo Pharmacetucials Inc. 185 Berry Street, San Francisco, CA 94107. Email: sking@napopharma.com.

Figure 1. Richard Evans Schultes conducting Dragendorff spot alkaloid field test in Colombian Amazon. Photo courtesy of R.E. Schultes.

studies in South America as part of the newly formed Institute of Economic Botany of the New York Botanical Garden.

After earning my degree in Biology, I worked for the National Academy of Sciences and then the Nature Conservancy. In 1989, Lisa A Conte, the founder of Shaman Pharmaceuticals, visited Dr. Schultes in Cambridge, Massachusetts, to learn about the feasibility of creating a pharmaceutical company that focused on ethnobotany as a source for discovering and developing new drugs. Dr. Schultes was highly supportive and shared with Lisa his enthusiasm for this project. He offered to be part of the Scientific Advisory Board of the company and directed her to many other ethnobotanists, ethnopharmacologists, and chemists, who shared his passion for the search for new drugs based on traditional ethnobotanical knowledge. He also indicated that such a company would need to have a full time ethnobotanist on the staff, not simply a person working on the outside as a consultant. Lisa visited Schultes several times and stayed in close contact with him as she sought to fund and create the company. Dr. Schultes also suggested Michael Balick as another key ethnobotanist to help her structure the company. Mike Balick also indicated it would be critical to have an ethnobotanist on the staff. I was identified by both Schultes and Balick as a good match for this work, and I was hired shortly after the first round of funding was obtained by Lisa Conte. Several of the first Scientific Strategy Team (SST) meetings took place in Cambridge, in order to be near Drs. Schultes and Raffauf, who were core scientists guiding the medicinal plant collection and analysis of these plants.

During one of the early SST meetings, a series of plants were identified as high priority research candidates, and one of those plants was *Croton lechleri* or Sangre de Drago. The red latex of this species was and is used widely in the Northwestern Amazon Basin to treat a number of conditions, including a topical application to heal wounds and orally to treat coughs, flu, ulcers and gastrointestinal disorders, including diarrhea (Ubillas et al. 1994). The company began focusing on the screening and development on this plant and several others from Africa and South East Asia.

In 1990, as these activities were getting underway, Dr. Schultes and Robert Raffauf published the book *The Healing Forest: Medicinal and Toxic Plants of the Northwest Amazon* (Schultes and Raffauf 1990). Dr. Raffauf was a friend and professor of pharmacognosy and medicinal chemistry at the College of Pharmacy and Allied Health Professionals at Northeastern University. Raffauf had spent 25 years in the pharmaceutical industry and had participated in several expeditions in the Amazon with Dr. Schultes and his students. The book contains information about 1500 species of plants used in traditional medicine from Brazil, Colombia, Ecuador, and Peru. These represented 596 genera and 145 plant families. At the time of the writing of the book, approximately 50% of the species mentioned had not been investigated for their chemical or pharmacological properties. *The Healing Forest* represented the fruits of close to 50 years of ethnobotanical fieldwork by Dr. Schultes. This was of course only a portion of the scientific focus of Dr. Schultes, and his other areas of research expertise have been described extensively in this volume and other publications.

Figure 2. Richard Evans Schultes interviewing Salvador Chindoy a key Kasma informant and colleague in the Sibundoy of Colombia. Photo courtesy of R.E. Schultes.

Dr. Schultes had been, for decades, urging the world to research, document, conserve, and respect the indigenous peoples and their knowledge of the tropical forests, long before it became a cause célèbre for international environmental organizations, human rights groups, and the concerned public. In the forward to the book *The Healing Forest*, Prince Philips noted:

> The plants and people of the Amazon alike are under severe pressure. Schultes and Raffauf emphasize the importance of ethno-

botanical conservation, and call attention to the fact that from these plants might come new chemical compounds of value to modern medicine and industry. It is their hope that the book will not only alert future generations of phytochemists to the potential of the Amazon as a source of new medicinal, toxic or other useful compounds, but in doing so will assist in the conservation of the folklore record of the indigenous peoples concerning this richly endowed region which is so critically important to the welfare of mankind.

Indeed Dr. Schultes devoted numerous publications to document ethnobotanical and ethnomedical knowledge. This included his own analysis of the records of early explorers of the Amazon for notes and clues indigenous ethnobotanical knowledge. A few of the titles to his article below provide glimpse of his passion for this topic.

- Ruiz as an Ethnopharmcologist in Peru and Chile (Schultes 1980)
- Several Unpublished Ethnobotanical Notes of Richard Spruce (Schultes 1985)
- Psychoactive Plants in Need of Chemical and Pharmacological Study (Schultes 1984)
- Burning the Library of Amazonia (Schultes 1994)
- Ethnnopharmacological Notes on the Flora of Northwestern South America (Schultes 1980)
- Notes, Primairly of Field Tests and Native Nomenclature, on Biodynamic Plants of the Northwest Amazon (Schultes 1983)

Research and development work at Shaman Pharmaceuticals continued throughout the entire decade of the 1990s on multiple plant extracts collected by teams of ethnobotanist and western trained medical doctors, who learned from indigenous men and women about plants used to treat infectious diseases, diabetes, and central nervous system disorders. The compound isolated from *Croton lechleri* SP-303, now known as crofelemer, continued through multiple phases of FDA-monitored clinical trials in the United States, Mexico, Venezuela, and Jamaica (Holodniy et al, 1999, DiCesare, et al 2002). The clinical indication for crofelemer became focused on diarrheal diseases, based in part on the ethnomedical usage of *Croton lechleri*.

The compound crofelemer as it turned out, was essentially not absorbed and this also made it an ideal compound to treat diarrheal diseases in the gut without having interactions with other drugs that may be absorbed. Research on the mechanism of action on crofelemer indicates that the compound appears to simultaneously target two distinct chloride channels. These channels, the cystic fibrosis transmembrane conductance channel regulator (CFTR) and the calcium activated chloride channel (CaCC) are located on the luminal membrane of the epithelial cells lining the intestines (Gabriel et al. 1999, Tradtranip et al. 2009). Crofelemer inhibits chloride secretion and reduces the gastrointestinal fluid accumulation and helps stop dehydration from excessive fluid loss.

In the time period from 2001, the year of Dr. Schultes' death, to the present in 2011, additional human clinical trials were conducted in the United States, India, and Bangladesh. Multiple studies demonstrated efficacy for the treatment of several diarrheal diseases, including infectious diarrhea and cholera-induced diarrhea (Crutchely et al. 2010). In November of 2010, the data for the final

Figure 3. Richard Evans Schultes examining his photographs at special exhibition at Stanford University in November 1992 with Lisa A. Conte. Photo courtesy of Shaman Pharmaceuticals Inc.

Phase 3 pivotal clinical trial, called ADVENT, were announced. Crofelemer demonstrated highly statistically significant results in the treatment of chronic diarrhea in people living with HIV/AIDS. The p-value achieved in this double-blind, placebo-controlled, two-stage, adaptive-design phase 3 study was 0.0096. Crofelemer was well tolerated, and the adverse event profile was not different from the placebo controlled group.

This positive Phase 3 data on a drug derived from ethnobotanical research on plants utilized by indigenous people of the Northwest Amazon will now lead to the filing of a New Drug Application (NDA) with the U.S. FDA in 2011 or 2012. In addition, a pediatric formulation of crofelemer is being developed, so that this non-absorbed anti-secretory first-in-class new chemical entity can be provided to children and communities where diarrheal disease still kill more then 1.5 million children per year. Crofelemer will become, along with oral rehydration solutions (ORS), zinc, cleaner water supplies, and increased sanitation, part of the global public health solution to stopping the huge rate loss of children's lives around the world. One of the primary goals of the founder of Shaman Pharmaceuticals Inc., Lisa Conte, this author, and the majority of the original Scientific Strategy Team (SST) members, including Dr. Schultes, was to be sure there would be and is global access to this novel, safe anti-diarrheal drug. This focus on global access will be accomplished by ex-

Figure 4. Portrait of Richard Evans Schultes at the opening of his Stanford University photographic exhibit, November 1992. Photo Courtesy of Shaman Pharmaceuticals Inc.

perts now working on a global access strategy for crofelemer, including Lady Neelam Sekhri Feachem of Napo Pharmaceuticals Inc.

One of the other core concerns of Dr. Schultes was the survival of both the tropical forest peoples and the tropical forest ecosystems. There is little disagreement that the peoples and communities who live in tropical forests share the similar aspirations for their families as the rest of the world. Those aspirations include access to health care, food, education, shelter, and the ability to provide a safe and secure long-term environment for their families. Due to the complex chemical structure of crofelemer, it cannot be made commercially via a synthetic process. The latex of this species is required for the isolation and purification of crofelemer. The latex of the *Croton lechleri* tree is sustainably harvested by local and indigenous communities and manufactured from this latex in collaboration with Napo Pharmaceuticals and its international partners. The sustainable harvesting process provides local people and communities with income generating activities, while they continue to live in and around primary and secondary forests.

The purpose of this brief paper is to pay proper homage and respect to one of the many visions, goals, passions, and accomplishments of Dr. Schultes. He was well aware of the drugs that had already been developed for the world pharmacopeia from plants used as medicines and poisons by the indigenous people of the New World. Dr. Schultes was tremendously supportive of Shaman Pharmaceuticals. He attended all of the Scientific Strategy Team meetings, provided numerous suggestions on plants we should study, and directed young and elder scientists to correspond with Shaman Pharmaceuticals on a regular basis. He also was quite outspoken if he had concerns about the scientific direction of the company and at times he expressed concern that the core ethnobotanical knowledge, which was the unique advantage of the ethnobotanical drug discovery process, was not being fully or adequately applied in the day-to-day operations. We always listened very carefully to his guidance, followed up on any and all plants leads that he suggested, and solicited his opinion on many aspects of the overall direction of the company. He wrote many very thoughtful and kind notes to me after our SST meetings, and expressed his empathy and compassion to Lisa Conte and me on several occasions. He was pleased when the non-profit organization created to return benefits from drugs developed was named The Healing Forest Conservancy. He was even more pleased when Ms. Katy Moran, a highly respected anthropologist, became the executive director of The Healing Forest Conservancy.

The development of this drug, crofelemer, soon to be finally approved by the FDA, is part of the legacy of Richard Evans Schultes. His wonderful personal manner of supporting so many people was extended to us all. A large percentage of the members of SST have been his students, directly or indirectly. These scientists, such as Michael Balick, shared Dr. Schultes' humility, sense of humor, respect for indigenous peoples, love for Amazonia and Colombia, in particular. All of us who had the good fortune and honor to work with Dr. Schultes are deeply grateful to him for his inspiration, discipline, and vision. He has been and continues to be in my heart each day.

## Acknowledgment

I would like to also pay homage to the indigenous people and communities that have collaborated with us on the discovery and development of crofelemer. I would also like to be sure to acknowledge Lisa Conte and the literally dozens of scientists who have contributed to the SST and the operations of the companies Shaman Pharmaceuticals and Napo Pharmaceuticals over the past 20 years. They also are due great thanks and appreciation for their work on crofelemer and the many other plants that have been and are part of the ethnobotanical drug discovery process.

## Literature Cited

Crutchley, R., J. Miller, and K. Garey. 2010. Crofelemer, a novel agent for treatment of secretory diarrhea. *Annals of Pharmacotherapy* 44.

Davis, W. 1997. *One River: Explorations and Discoveries in the Amazon Rain Forest.* Simon & Schuster, New York.

———. 2004. *The Lost Amazon: The Photographic Journey of Richard Evans Schultes.* Chronicle Books, San Francisco.

DiCesare, D., H.L. DuPont, J.J. Mathewson, et al. 2002. A double blind, randomized, placebo-controlled study of SP-303 (Provir) in the symptomatic treatment of acute diarrhea among travelers to Jamaica and Mexico. *Am J Gastroenterol* 97:2585–2588.

Gabriel, S.E., S.E. Davenport, R.J. Steagall, V. Vimal, T. Carlson, and E.J. Rozhon. 1999. A novel plant-derived inhibitor of cAMP-mediated fluid and chloride secretion. *Am J Physiol* 276:G58–G63.

Schultes, R.E. 1980. Ruiz as an Ethnopharmcologist in Peru and Chile. *Botanical Museum Leaflets* 28(1):87–122.

———. 1980. Ethnnopharmacological Notes on the Flora of Northwestern South America. *Botanical Museum Leaflets* 28(1).

———. 1983. Notes, Primarily of Field Tests and Native Nomenclature on Biodynamic Plants of the Northwest Amazon. *Botanical Museum Leaflets* 29(3):251–272.

———. 1984. Psychoactive Plants in Need of Chemical and Pharmacological Study. *Proc Indian Acad Sci (Plant Sci)* 93(3):281–304.

———. 1985. Several Unpublished Ethnobotanical Notes of Richard Spruce. *RHODORA* 87(851):439–441.

———. 1994. Burning the Library of Amazonia. *The Sciences* March/April:24–30.

Schultes, R.E., and R.F. Raffauf. 1990. *The Healing Forest: Medicinal and Toxic Plants of the Northwest Amazon.* Dioscorides Press, Portland, Oregon.

Tradtrantip, L., W. Namkung, and A.S. Verkman. 2010. Crofelemer, an Antisecretory Antidiarrheal Proanthocyanidin Oligomer Extracted from *Croton lechleri*, Targets Two Distinct Intestinal Chloride Channels. *Molecular Pharmacol* 77:69–78.

Ubillas, R., S.D. Jolad, R.C. Bruening, M.R. Kernan, S.R. King, D.F. Sesin, M. Barrett, C.A. Stoddart, T. Flaster, J. Kuo, F. Ayala, D. Smee, R. Sidwell, K. Soike, A. Brazier, S. Safrin, R. Orlando, P.T. Kenny, N. Berova, and K. Nakanishi. 1994. SP-303, an antiviral oligomeric proanthocyanidin from the latex of *Croton lechleri* (sangre de drago). *Phytomedicine* 1:77–106.

# The Leaf Falls near the Tree: My Father and My Career in Plant Biology

Neil P. Schultes[1]

I find myself in the unique position of being not only one of Richard Evans Schultes many students (as an undergraduate in his *Plants and Human Affairs* course) but also a recipient of half his genome. Is it the *nature* of a Schultes or the *nurture* in the Schultes household that led to my career choice as a plant biologist? One might think that as my father and I share similar fates—attending undergraduate and graduate school at Harvard and choosing plant science as a career—that there might have been a hefty dose of generational grooming. Surprisingly, this was not the case. What my father offered me, as well as to many others, were opportunities and experiences. These undoubtedly helped me decide upon my career. Today I work at The Connecticut Agricultural Experiment Station in New Haven CT as a plant molecular biologist. There, colleagues and I work as part of a large effort in the plant community to identify and understand how maize genes give rise to C4 photosynthesis. Despite the gap between field botany and lab research it never ceases to amaze me how often I reach back and retrieve informational gems from his course to use in my own seminars and teaching. Truly his legacy extends beyond the *nature* of Economic Botany and into the *nurturing* of the next generation of plant biology students.

Naturally I was honored when asked to deliver reminisces of my father at Botany 2011 meeting, and quite frankly felt astonished to be included among the panel of his former students. I would like to thank all who worked so hard putting the symposium together, and for inviting and hosting me as well. I find myself in the unique position of being not only one of Richard Evans Schultes (RES) many students (although merely an undergraduate in his *Plants and Human Affairs* course) but also a recipient of half his genome. As such, I have inherited a portion of his dry sense of humor, which many of you are well acquainted with, and ask forgiveness in advance if my tone is informal and tongue and cheek.

## Nature or Nurture

As we are all well aware, my father was a field botanist and a Harvard man through and through. Similarly, I received all my degrees through Harvard and am now a plant molecular biologist working in the Department of Biochemistry & Genetics at The Connecticut Agricultural Experiment Station in New Haven, Connecticut. The inevitable age old question is obvious. Is it the nature of a Schultes or the nurture in the Schultes household that led to my career choice as a plant biologist? One might think that as my father and I share similar fates that there might have been a hefty dose of generational grooming. Surprisingly, this was not the case. What my father offered me, as well as to many others, were opportunities and experiences that helped shape future decisions. Now I will give away the punch line. The take home message for my talk is a mutated form of a common trite saying, "It's the journey AND the destination." What Dad realized and experienced in his life is that your education is just as much formed by the process (the journey) as by the end result of your endeavors (the destination). These "journeys" undoubtedly helped

[1] Department of Biochemistry & Genetics, The Connecticut Agricultural Experiment Station, 123 Huntington St, New Haven, CT. Email: Neil.Schultes@ct.gov.

me decide upon my career. One "journey" I experienced left its mark on me and countless others as well. I refer to the extraordinary course that dad taught for many years, *Plants and Human Affairs*. Despite the gap between taxonomy based botany of Biology 104 and my current laboratory research, it never ceases to amaze me how often I reach back and retrieve informational gems from his course to use in my own seminars and teaching. Truly his legacy extends beyond the nature of Economic Botany and into the *nurturing* of the next generation of plant biology students.

## PH.D. GRADUATION

My father and I share many phenotypic similarities that are tantalizing for a geneticist to observe, but sadly are by themselves inadequate to answer the nature/nurture question. Nonetheless, they give a view into who the Schultes' are. In what was one of my most memorable days, my father and I are standing outside the Botanical Museum on Oxford Street in Cambridge festooned in our finest crimson plumage (Fig. 1). It was commencement day, 1990, the year I received my Ph.D. in Genetics from Harvard University.

Now aside from the similarities in appearance, what is unseen in this photograph are two people fully committed to studying plants albeit from different ends of the plant biology research spectrum. For both, our early years at Harvard offered us experiences that forever shaped our destiny and led us down the "path less taken." In addition, both my father and I share the view that we are indeed lucky to spend our lives combining career and passion—few have that opportunity. Now, perhaps many of you do not know that Dad *lived* for this day every year. Harvard

Figure 1. Harvard commencement day, 1990. Photo courtesy of the author.

Commencement was a sacred day that he was wont to miss. We see here that he is carrying a baton and a name tag designating him as a Marshall. I asked him why be a Marshall? He smiled and in his dry dead pan voice said it was the only time of the year he could push Deans around!

## SCHULTES GENERATION TIME

There are other similarities too in the Schultes family. As a geneticist, I am used to working with "model" organisms, such as *Escherichia coli*, *Saccharomyces cerevisiae*, and *Arabidopsis thaliana*, that have relatively quick generation times. Not so with the Schultes clan. Over three generations the Schultes clan averages 40 years between the generations! My grandfather, Otto Schultes, was born in 1875, siring his first child, Richard, at the age of 40 years. Pictured here (Fig. 2) is perhaps the earliest photograph of RES resting on his father's knee during the summer of 1915. At the age of forty-five, Dad married Dorothy McNeil, and they soon formed a family of my older brother, Richard Evans, my twin sister, Alexandra Ames, and myself. In true tradition, I followed suit with my wife, Martha Howard, having our daughter, Olivia, at ages 35 and our son, Evans, at 40 years. In half jest one can ask, "Was this predestined or arising from tradition?" We may never know.

Figure 2. Otto and Richard Evans Schultes ca. 1915. Photo courtesy of the author.

## Plant Processing Locus

I offer one last tongue and cheek commonality among the Schultes males. Even though the human genome has been sequenced and fully annotated, there might just be another gene lurking in the Schultes genome that has heretofore been overlooked. I have dubbed it the Plant Processing locus or *PPL*. One might assume that, along with dad's incredible love for botany, he would be a skilled horticulturalist. Quite the contrary, as far as indoor plants are concerned, my father (and I) have little aptitude or skill in successful plant rearing and even less interest. The only exception was the peyote plant that he brought back from one of his trips when I was twelve. He presented it to me after I had written a school report on peyote earlier that year. The plant flourished in our house for many years. Now normally my father took plants, soaked them in organic solutions, then crushed them in a plant press. My *PPL* has a slightly different phenotype, wherein I first crush plants with a mortar and pestle then extract the nucleic acids with organic solvents! Recently, this phenotype has appeared in my son Evans, who for a third grade science project made a plant press and described what plant explorers do. Perhaps there will be a third generation of Schultes plant biologists! Now that would make Dad very happy indeed.

## It's the Journey and Destination

My father was a product of his times, growing up in a straight-laced Victorian-era middle class family in East Boston. So another question arises: how did my father break away from his background and pursue quite an unconventional career and life? He has often been regarded as one of the "last Victorian explorers." That is hardly what good boys do from East Boston—at least according to his mother, Maude, who wanted him to become a physician. Frankly, his upbringing was from the end of the Victorian era. We must keep in perspective just how different his formative years were from the present. Here are just a few examples. Consider that while attending East Boston High School he took French,

Figure 3. RES wearing his East Boston High School Marching Uniform 1932. Photo courtesy of the author.

Latin, and Greek! These studies are hallmarks of a gentleman's training, left over from decades earlier view of what a proper liberal arts education should entail. How many high schools today teach Greek? Gymnastics class consisted of marching. Here he is seen at age seventeen in his East Boston High School Marching uniform (Fig. 3). Apparently physical education as we know it hadn't quite made it to East Boston in the 1930s. (Interesting my brother, Evans, resurrected this very coat from the basement in the early 1970s and proudly wore it as part of counterculture garb, showing that fashions are indeed recycled!) Who would have though that someone from this background would have led such a life?

Once he entered Harvard, my father experienced a series of life changing "journeys" that altered his destiny. Foremost was his enrollment in the *Plants and Human Affairs* course offered by his professor and mentor, Oaks Ames. Most important was his famous peyote *trip*—and here I mean the physical not the chemical one—he took as an undergraduate with Weston La Barre, a graduate student from Yale, to live among the Kiowa in Oklahoma. There he had his epiphany and the literal journey *and* the destination (culmination in his undergraduate thesis on peyote) were certainly pivotal events that shaped his life. For me, the trait I most respect in my father was his ability to learn from his experiences, recognize their uniqueness and importance, and incorporate them into his life. This is the legacy that he passed on to his students and to us.

I will not dwell on the many other "journeys" that shaped father's life, as these have been well documented and told particularly by Wade Davis in his wonderful book *One River*. To him, I am forever indebted. I have just one photograph that I ran across among father's collection that illustrates his keen observation even for the unexpected, his attention to detail and making the most of an experience. Here we see a young RES inspecting the flower of a plant, perhaps new to him, on his travels (Fig. 4A). On the reverse side of the picture we see in his meticulous handwriting, "Examining and extremely rare endemic plant collected only once before by von Martius in 1820. Schoenobilus Araracuara Rió Caquitá, Colombian Amazonas" (Fig. 4B). It is clear that he noticed this unique and rare plant and later spent the time to investigate this rare find. For me, this is fine example of how this small "journey" was followed through to its "destination," thereby enriching his skill, interest, and passion. These are the lessons and skills that he passed on to the next generation.

Figure 4A. RES examining a rare plant. Photo courtesy of the author.

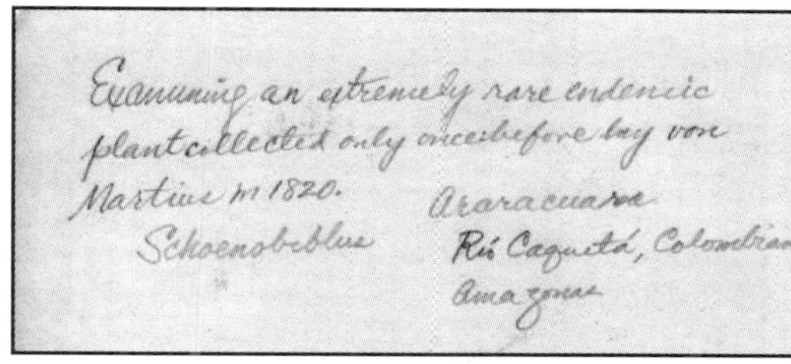


Figure 4B. RES meticulous handwriting on the back of photo. Photo courtesy of the author.

## My Time in the Amazon

I had the amazing opportunity to travel to the Colombian jungle with my father. Imagine being a fourteen year old in the Vaupes! We traveled along with Jim Henderson, a friend from school, and a number of my father's students to Mitu. I had a number of small adventures, including too close associations with local wasps! Earlier in the symposium, Jim Zarucchi, who also accompanied us on the trip, related some other "experiences" with local plant products! For me, the trip did not mark my epiphany and turn my mind toward a career in plant science. Nonetheless, it did make a lasting impact on my memories. I have been carrying a now very ragged photo in my wallet since 1975 showing our group collecting specimens by the river; it is a nice reminder of special times past (Fig. 5). My epiphany or "journey," which convinced me to pursue plant biology, was during my time as an undergraduate in the laboratory of Dr. Lawrence Bogorad in the Biological laboratories directly behind the Botanical Museum at Harvard.

## What I Do Now

Today, I work in the Department of Biochemistry & Genetics at The Connecticut Agricultural Experiment Station in New Haven, Connecticut, as a plant molecular biologist. There, my colleagues, Drs. Richard Peterson and Israel Zelitch, and I investigate the genetics and physiology of how maize activates C4 photosynthesis in photosynthetic tissues. This is part of a much larger effort in the plant community

Figure 5. The author at 14 collecting specimens near Mitu. Photo courtesy of the author.

to identify maize genes that control C4 photosynthesis through a systems biology approach. Our corner of this project is to employ advanced photo-physiological measurements to developing maize leaves under normal conditions and conditions where the photorespiration is compromised. These data will be used in conjunction with transcriptomic and proteomic data to help identify master regulatory genes in photosynthesis. The driving force behind these investigations is the hope that rice, which now performs inefficient C3 photosynthesis, can be altered to perform more efficient C4 photosynthesis, thereby eventually paving the way for increased harvest yields. Although far removed from field botany, my father was nonetheless proud of my career choice, and I in turn was grateful for his offering opportunities throughout my life that helped me find my passion.

### Small World Coincidences

I will end my paper with an example of how it is indeed a small world. My father would have been highly amused at the connections between where I work now and Harvard. Before I took my position at The Experiment Station, I was unaware of these connections, so they did not influence me in my job decision. In earlier days, the Department of Biochemistry & Genetics at The Connecticut Agricultural Experiment Station was made up of separate departments of Biochemistry and Genetics. Donald Jones, a member of the Genetics department, was instrumental in developing the genetics of double hybrid corn, a technique that had profound effects on the world. One of his graduate students was Paul Mangelsdorf, who received his degree from Harvard but did most of the corn genetics at The Experiment Station's Lockwood Farm with Jones.

Jones and Mangelsdorf continued their collaborations, developing the genetic system for including male sterile plants in hybrid seed production. Again their work—in its day—represented a huge advance for agriculture. Dr. Mangelsdorf was, of course, the director of the Harvard Botanical Museum, my father's boss, and a great supporter and mentor of my father during his many years in the Amazon. Surprisingly, there is a direct decent from my time at Harvard and the Biochemistry department. My Ph.D. professor, Jack Szostak at the Department of Genetics in Harvard Medical School, was a student of the plant geneticist Ray Wu, who in turn trained under Wright Wilson. Wilson's mentor was Lafayette Mendel at Yale University. Mendel and Dr. Thomas Osborne in the Department of Biochemistry at The Experiment Station worked together for many years leading to the discovery of Vitamin A. I think that many of us can trace our linage to Harvard and my father, if not in direct descent certainly through his influence. Again my father's legacy truly extends beyond the *nature* of Economic Botany and into the *nurturing* of the next generation of plant biology students.

# Medicinal Plants and the Legacy of Richard Evans Schultes

Djaja D. Soejarto[1]

The late Richard Evans Schultes was a complex scientist. He was an accomplished ethnobotanist, plant taxonomist and economic botanist, plant conservation and ethnobotanical conservation scientist, ethnopharmacologist, and a medicinal plant explorer. The author has had the privilege to be a student, colleague, and friend of Dick Schultes and inherited his teachings, in the following order of importance: medicinal plant explorations, taxonomic study of *Saurauia* (Actinidiaceae), and ethnobotany. Schultes' legacy lives on as demonstrated by research and teaching carried out and accomplished by this author in the fields of medicinal plant explorations and research, studies on the taxonomy of Actinidiaceae, and research and teaching in ethnobotany.

Richard Schultes was a complex scientist. Many words have been used to describe him and his scientific expertise. He was first and foremost an ethnobotanist—in fact, he is The Father of Modern Ethnobotany. He was also a plant taxonomist, an economic botanist, a conservation scientist, and an ethnopharmacologist. Schultes' images are all over the Internet and speak louder than words about his multiple, complex expertise (Schultes Images 2011).

As the Father of Ethnobotany, his fatherly images also speak for themselves, painting close relationships between him and the people he worked with and studied. In his interview with Margaret Kreig, Schultes made this statement about his jungle collaborators: "Much can be accomplished if the ethnobotanical investigator treats natives as a gentleman should. He must realize that, far from being a superior individual, he—the civilized man—is in many respects far inferior to the native in the native's own environment" (Kreig 1964, p. 89). Later, in an article by E.J. Kahn, Jr. (1992), he is quoted to have said: "I do not believe in hostile Indians. All that is required to bring out their gentlemanliness is reciprocal gentlemanliness."

Richard Schultes was an accomplished plant taxonomist and economic botanist. His writings on the taxonomy of *Hevea* (Euphorbiaceae) (Schultes 1947, 1949, 1955, 1960, 1978), and his publications of numerous new species of *Micrandra* (Euphorbiaceae) and *Saurauia* (Actinidiaceae) (Tropicos 2011), among many others, are testament to his accomplishments in these fields. He also made more than 24,000 collections of plants in his lifetime. He has been memorialized in the plant scientific names of many plant genera and species. Examples are a Solanaceous genus *Schultesianthus* Hunz., a Cyclanthaceous genus *Schultesiophytum* Harling, and a Gesneriaceous genus *Resia* H.E. Moore. More than 60 plant species bear his name under the specific epithets *schultesii*, *schultesianus*, and *schultesiana* (Prance 2001). One of Schultes' most treasured achievements in the field of botany is the gold medal award he received from the Linnean Society of London in 1992. This is a most prestigious award, regarded as a sort of Nobel Peace Prize in Botany (Harvard University Herbaria 2011).

Dick Schultes was also an accomplished plant conservation scientist. The 1984 WWF-Duke

[1] Program for Collaborative Research in the Pharmaceutical Sciences, Department of Medicinal Chemistry and Pharmacognosy, College of Pharmacy, University of Illinois at Chicago, 833 S. Wood St., Chicago, IL 60612 and Department of Botany, Field Museum, 1400 S. Lake Shore Dr., Chicago, IL 60605. Email: dds@uic.edu.

of Edinburgh Conservation Medal (The Duke of Edinburgh Conservation Medal 2011) is one proof of this accomplishment. This is the premier award recognizing highly meritorious contributions to the conservation of wildlife and natural resources. In the same year, he was also presented with the Tyler Award for his Environmental Achievement, and in 1991, he was presented with the Lindbergh Award. On September 1, 1986, the President of Colombia dedicated a 2.2 million hectares of forest located in La Macarena in eastern Colombia as "Sector Schultes" (Harvard University Herbaria, 2011). His writings on the role of ethnobotany in biodiversity conservation (Schultes 1986, 1991, 1994a) made Schultes a champion of ethnobotanical conservation.

## Richard Evans Schultes: Medicinal Plant Explorer

From the start of his study of the peyote plant as an undergraduate thesis at Harvard University, Schultes' future career as an ethnobotanist became intertwined with ethnopharmacology. This dual expertise was strengthened and became prominent with his subsequent studies of teonanacatl, *Virola* spp., ololiuqui, and other hallucinogenic and narcotic plants, in collaboration with pharmacologists, physicians, and chemists.

Early in 1941, when he traveled to the Amazon to study curare under a grant from the National Academy of Sciences, the focus of his ethnopharmacological study shifted into a field of endeavor today recognized as bioprospecting (Davis, 1996).

As is widely accounted in numerous published biographical sketches and articles, and obituaries, and in Wade Davis' book, *One River*, Dick Schultes returned from the depth of the Amazon to Bogota on December 21, 1941, two weeks after Pearl Harbor was attacked by the Japanese on December 7, 1941, to report to the U.S. Embassy and to enlist for military service. When the U.S. government asked Schultes on November 20, 1942, to return to the Amazon to search for rubber, from any plant source, particularly from the *Hevea* spp., to serve as a substitute of the *Hevea brasiliensis* rubber plantation in Southeast Asia lost to the Japanese, instead of sending him to the war, the stage was set for his bioprospecting explorations. He collected more than half-a-million seeds of the rubber plant (*Hevea brasiliensis*) from Leticia alone in 1945, which would become the germplasm of experimental rubber plantation in the New World (Davis, 1996).

At a later date, when Dick Schultes continued his studies on psychoactive plants after his return from the Amazon, his bioprospecting passion was directed to medicinal plants (Schultes 1972) and is reflected in the following assertion: "Nowhere perhaps have the potentialities of ethnobotanical investigation been more scintillating than in the search for new psychotomimetic drugs" (Schultes 1967). Jointly with a chemist, the late Tony Swain, Schultes declared that "The plant kingdom is a virgin field for biodynamic constituents" (Schultes and Swain 1976). Still, at a later date, Dick Schultes made this statement: "During the past 5 decades, I have conducted ethnopharmacological research amongst Indians in the northwest Amazon, especially Colombia. Of the nearly 1500 species of plants in 596 genera and 145 families that I have recorded as medicines or poisons, there are 25 species in 15 families valued by the natives in treating, usually amongst the elderly, conditions that superficially appear to resemble Alzheimer's disease or related mental problems. There is, to be sure, no reason to suppose that any or all contain cholinesterase-inhibiting compound. But, in view of the current interest in Alzheimer's disease, it might be worthwhile to subject some of the plants that I shall mention below to technical examination" (Schultes 1993).

He also asserted that, "If chemists expect to obtain by random collecting sufficient quantities of 80,000 species and all their ecotypes from this area of such difficult travel and transportation, the task may never be completed. It would be more productive to concentrate on those plants that the indigenous peoples have found, over millennia of experimentation, to be bioactive and which they have used in their medicine as ameliorators, stimulants or curative or psychotic agents." These, to Dick Schultes, represent "... an incredible chemical laboratory still practically untouched by scientific investigation for new therapeutic agents ...." (Schultes 1994b).

## THE LEGACY THAT I INHERITED FROM RICHARD EVANS SCHULTES

The first legacy that I inherited from Dick Schultes is an expertise in medicinal plant explorations and studies embodied in several plant collection expeditions, which he either delegated me to run, or on which he took me as a member of his expeditions, in 1963, 1964, 1965, 1966, while I was a graduate student at Harvard University. The second legacy is a taxonomic expertise, resulting in the taxonomic revision of South American *Saurauia* (Actinidiaceae), the focus of my Ph.D. dissertation. The third legacy is the study of ethnobotany.

On the medicinal plant explorations, the first expedition was plant collection work in the Department of Cundinamarca of Colombia in the summer of 1963. At that time, Schultes was teaching an Economic Botany course at the Instituto de Ciencias Naturales, Universidad Nacional de Colombia [Institute of Natural Sciences, National University of Colombia] in Bogota, Colombia, while I was assigned to carry out fieldwork to collect plant samples for chemical analysis at the Research Triangle Institute (RTI) in North Carolina.

Immediately after our arrival in Bogota from Cambridge, MA, before the start of the Bogota course, he gave me an on-site training in plant sample collection methods. He also introduced me to the plants in their natural habitat, which would become the subject of my Ph.D. research, the genus *Saurauia* (Actinidiaceae). The Cundinamarca expedition was completed at the end of August, at the same time that Dick Schultes completed his course. On September 1, we traveled to Leticia (Amazonas) to explore and to continue the plant collection work in the forests surrounding Leticia. We also traveled to a place called Puerto Nariño, about 80 km upstream from Leticia (Fig. 1). By mid-September we finished our work in Puerto Nariño, and Dick returned to Leticia on a motorboat. He left me to canoe back to Leticia, while also collecting plants along the way. By the end of the summer, a total of 68 plant samples were collected and shipped to RTI.

In the summer of 1964, Dick Schultes sent me back to Colombia to lead a full-fledged plant collecting expedition in southern Colombia. Professor Hubert Vogelmann of the University of Vermont joined the expedition together with a graduate student at Harvard, Fred Olday. We undertook collection work in the Departments of Nariño and Putumayo and collected 67 samples belonging to 67 species of

Figure 1. Schultes, Soejarto and collaborators at the mouth of Río Loretoyacu, Puerto Narino, Amazonas, September, 1963. This photograph was taken under the tree of *Ormosia amazonica* Ducke var. *venenifera* R.E. Schultes, based on voucher herbarium specimen *Schultes, Raffauf & Soejarto 24107* (COL, GH, US). Photo courtesy of the author.

angiosperms. All samples were dispatched to RTI. Following the conclusion of this expedition, I concentrated my effort to carry out field research on my plants, the genus *Saurauia*, in the Departments of Nariño and Putumayo in the southern part of Colombia.

In the summer of 1965, I was sent by Dick Schultes to the Amazon of Colombia to recollect about 300 kg of plant samples of *Isertia hypoleuca* Benth. (Rubiaceae), which he and I had collected (*Soejarto & Schultes 746*) in the Amazon during our 1963 expedition. Samples of this plant turned out to show good results in chemical analysis at RTI, so a larger amount of materials was requested. I should note that I was carrying out this particular [1965] expedition with a student at Harvard Medical School, the now-renowned Andrew Weil (Weil 2011).

Once we completed our collection of *Isertia hypoleuca* and dispatched the collected materials to RTI, Andy and I traveled together all the way south to La Paz, Bolivia. We returned by land to Pasto, the capital city of the Department of Nariño (Colombia), through Lima (Peru), Guayaquil (Ecuador), Quito, and Ipiales (a border city between Colombia and Ecuador). We collected plants along the way, including *Saurauia* specimens. We split in Pasto. Andy returned to Bogota, and the U.S., while

I stayed to carry out field studies on *Saurauia* in the Departments of Nariño and Putumayo.

In the fall of 1966, Dick Schultes together with the late Robert Raffauf, a chemist and professor at Northeastern University, Boston, ran an expedition to the Department of Vaupés in the Colombian Amazon. A botanist at the Colombian Institute of Natural Sciences, Bogota, Enrique Forero, and myself, joined the expedition as field assistants. Enrique Forero stayed in the expedition for only a short time. We focused our collection work in the forests along the tributary of the Vaupés River in Mitú, and in a location called Kubiyu, a half-a-day trip upstream by a motorboat. One important innovation of this expedition was the actual field-testing of fresh plant parts to detect the presence of alkaloids in the plants, using a modified Dragendorf reagent. Plant samples were collected in the event of active tests, but herbarium voucher specimens were collected regardless of active or non-active test results. A few of the plants collected were the results of an ethnobotanical interview process with local populations. The results of this expedition served as the platform for Dick Schultes' book, *The Healing Forest*, co-authored with Bob Raffauf (Schultes and Raffauff 1990). Some of the plants we collected are cited in this book under the SRS (Schultes-Raffauf-Soejarto) collection number.

## THE LEGACY OF RICHARD EVANS SCHULTES THAT I CARRIED ON INTO MY ACADEMIC CAREER

### Solidifying and Continuing the Legacy

I completed my doctoral dissertation, "Studies of South American *Saurauia* (Actinidiaceae)" (Soejarto 1968, 1980), at Harvard University and received my Ph.D. degree in Biology in 1968. Because of my good knowledge of the Spanish language and my extensive fieldwork experiences in Colombia, Ecuador, and Peru, immediately after my graduation, Professor Schultes suggested that I should apply for a fellowship program at Tufts University. In 1969, I was awarded a fellowship by the Latin American Teaching Fellowships Program at the School of Law and Diplomacy, Tufts University, and was assigned to serve as a Latin American Teaching Fellow at the Department of Biology, University of Antioquia, Medellin, Colombia. As an LATF (1969–1971), I founded the Herbarium of the University of Antioquia [HUA] on October 24, 1969 (Corréa 1994), and trained students in plant taxonomy and economic botany. My former students, and later their students ("academic grandsons") are presently running this herbarium institution, the second largest in Colombia, which houses 180,000 plant collections (Index Herbariorum 2011), mostly from northwestern Colombia. Following the completion of the LATF term, I joined the Department of Biology as a faculty member.

While on the faculty at the Department of Biology of the University of Antioquia, I received a letter in 1974 from Dick Schultes, sounding me out about whether I would be interested to work on a literature-based project on plants used for fertility regulation. My answer was "yes," and I was appointed as a Research Associate in Ethnopharmacology at the Harvard Botanical Museum for the period of 1974–1975. My mission was to compile information on plants for fertility regulation based on literature search. The report I produced came to the hands of the managers of the Task Force on Plants for Fertility Regulation of the World Health Organization

(WHO), Geneva. This was the beginning of my association with WHO (1977–1979) as a botanist-consultant, and a point of acquaintance and association (Soejarto et al. 1978) with the late Prof. Norman R. Farnsworth, Distinguished University Professor of Pharmacognosy at the University of Illinois at Chicago, and Distinguished Economic Botanist (1983) (Farnsworth 1984), and Director of the then Department of Pharmacognosy and Pharmacology at the UIC College of Pharmacy. At the end of my term with WHO, I joined the Department of Pharmacognosy and Pharmacology, of which Norman Farnsworth was Director. Initially, I served as his postdoctoral fellow, becoming a member of his drug discovery and development team, including the development of a natural products database, now online and globally recognized as the premier source of information on the ethnobotany, chemistry, and pharmacology of bioactive compounds derived from plants (Farnsworth et al. 1981, NAPRALERT 2011). Thanks to Norman Farnsworth's mentoring, I solidified my expertise in medicinal plant studies and explorations, within the setting of a multidisciplinary, collaborative research on the discovery of new bioactive compounds from plants as potential candidates for development into medicines.

My career in medicinal plant explorations is highlighted by a series of contract awards from the U.S. National Cancer Institute, NIH (1986–2004), and the ICBG grants from the Fogarty International Center, NIH, 1998–present.

## Exploring for Medicinal Plants in Southeast Asia

The NCI contracts had a mission to supply plant samples collected from the tropical rainforests of Southeast Asia for anticancer and anti-HIV testing at the laboratories of the NCI, Frederick, MD. This exploration program was a collaborative endeavor between the UIC, the Arnold Arboretum of Harvard University, and the Bishop Museum, Hawaii, for which I served as the Principal Investigator (Soejarto 1993, Soejarto et al. 1996) (Fig. 2). Although the administrative base of this project in Chicago was at the University of Illinois at Chicago, the research base was at the Department of Botany of the Field Museum, where I have held a Research Associate appointment since 1977. A number of institutions (U.S.-based and South East Asia-based) and many botanists, U.S.-, and foreign-based, collaborated in this plant collection program. At the time of the termination of the contract in 2004, 21,941 plant samples, comprising approximately 4,839 species of angiosperms, in 1,349 genera and in 271 families, were collected and submitted to the NCI Repository at Frederick, MD.

A major contribution that resulted from this NCI-sponsored exploration program was the discovery by the NCI of the anti-HIV compounds, the calanolides, (+)-Calanolide A and (-)-Calanolide B (also called costatolide), from *Calophyllum lanigerum* Miq. var. *austrocoriaceum* (Whitmore) P.F. Stevens and *Calophyllum teysmannii* Miq. var. *inophylloide* (King) P.F. Stevens (Clusiaceae), respectively (Kashman et al. 1992). Both plants were collected in Sarawak, Malaysia, and both compounds have 100% kill on HIV, as potential antiretroviral drug (reverse-transcriptase inhibitors) (Kashman et al. 1992, Fuller et al. 1994, Boyd et al. 1997). Calanolide B was found in abundance in the latex of *C. teysmanni inophylloide*, up to 66% of yield of the latex extractive (Lin et al. 1999). Furthermore, the source material

Figure 2. Soejarto at work in the evening to document plants collected for anticancer screening during a plant collection expedition in Palawan, Philippines, 1988, under the sponsorship of the U.S. National Cancer Institute. Photo courtesy of the author.

(latex) may be harvested sustainably (Boyd et al., 1997, Soejarto et al. 1997). In June 1992, (+)-Calanolide A was approved by a Decision Network Meeting at NCI for development into a drug to treat AIDS, while Costatolide was approved on May, 1993, as an alternative candidate. In 1995, the compound (+)-Calanolide A was licensed to an Illinois-based pharmaceutical company, MediChem Research, and in 1996, this company successfully developed a method to produce Calanolide A through total synthesis (Flavin et al. 1996). In the same year, the company established a daughter company called Sarawak MediChem, a joint venture between MediChem Research and the State Government of Sarawak (Sarawak Medichem Pharmaceuticals, Inc. 2011). By 2005, clinical trials Phase I-a and I-b were completed and clinical trials Phase II-a was being scheduled (Highbeam Business 2011, Hot Stocked 2011).

In 2006, ownership of Sarawak-Medichem was transferred to the State Government of Sarawak, and as of December 20, 2006, the company is a subsidiary of Craun Research Sdn. Bhd. (Bloomberg Businesweek 2011). The latest news on the development of the calanolides came out in 2009, when the Deputy Chief Minister of Sarawak declared that "Sarawak is closer to producing anti-HIV drug" (The Star Online 2011).

Aside from the development of potentially important drugs to treat AIDS, the discovery of the calanolides led to an important turning point in the biodiversity conservation and bioprospecting laws of the government of

Sarawak. To protect the calanolide-producing plants and the biodiversity of the country, in June, 1993, the government enacted an ordinance, Calophyllum Species Order (prohibition of felling and export) under Section 96(a) and (b) of the Forests Ordinance (The Calophyllum Story 2011). This Calophyllum Species Order led to the enactment of the Sarawak Biodiversity Centre Ordinance in 1997, and eventually, the establishment of the Sarawak Biodiversity Centre (SBC) the following year (1998), whose mission is "... to initiate programmes for the conservation, utilization, protection and sustainable development of biodiversity in the State" (Sarawak Biodiversity Centre 2011).

## Discovery of Bioactive Compounds from Plants of Vietnam and Laos

In 1998, while the NCI contract was still ongoing, a multidisciplinary team of botanists, ethnobotanists, natural product chemists, biochemist, economists, biotechnologists, and an anthropologist and attorney, under my leadership as Principal Investigator, was awarded an ICBG Grant to study the biodiversity of Vietnam and Laos. The primary goal of the project was the discovery of new bioactive compounds from plants of Vietnam (Cuc Phuong National Park, Nho Quan District, Ninh Binh Province) and Laos (throughout the country), as potential candidates for pharmaceutical development for therapies against cancer, AIDS, tuberculosis, and malaria. Both the "random collection" approach [namely, a collection program that maximizes taxonomic diversity] and the ethnobotanical approach were used in plant sample acquisition as basis for discovery. The second goal was the promotion of biodiversity conservation in Vietnam and Laos, while the third goal was the promotion of economic development among communities that participated in the ICBG project in both countries, as well as at the ICBG host institutions in Laos and Vietnam (Soejarto et al. 1999, 2006). The operation of this ICBG consortium is governed by a 6-way Memorandum of Agreement, that includes terms on benefit-sharing (Soejarto et al. 2004a, 2007).

The results of the medicinal plant explorations effort (including Vietnam and Laos) of this ICBG may be summarized as follows. By July, 2008, a total of 4,691 plant samples had been collected through the "random" collection approach from the Cuc Phuong National Park in Vietnam, comprising 1,600 species of angiosperms, while 1,188 plant samples, comprising ca. 900 species of angiosperms, had been collected through interviews with healers in Laos (Soejarto et al. 2009, 2012), and with healers in communities surrounding Cuc Phuong National Park. These samples were extracted using a standardized protocol, then the extracts were tested against cancer, HIV, tuberculosis, and malaria disease systems.

As of October, 2010, 2,786 extracts had been tested (in vitro) against 7 cancer cell lines, resulting in the isolation of 84 compounds, of which 33 were new, 37 active, and 15 new and active. For AIDS, 3,507 extracts had been tested against HIV, resulting in the isolation of 112 compounds, of which 39 were new, 50 active, and 26 new and active. For tuberculosis, 2,006 extracts had been tested against *Mycobacterium tuberculosis* (virulent strain), resulting in the isolation of 42 compounds, of which 7 were new, 11 active, and 4 new and active. For malaria, 2,268 extracts had been tested against *Plasmodium falciparum* parasites (chloroquine-sensitive and chloroquine-resistant clones), resulting in the isolation of 76 compounds, of which 23 were new, 29 active, and 12 new and active. Overall, a total of 314 compounds had been isolated, represented by 102 new, 127 active, and 57 new and active compounds.

As of this date, our team continues to work on the pre-clinical development of 6 potent anticancer and anti-HIV compounds isolated from plants collected in Cuc Phuong National Park. Provisional patents have been filed for both anticancer and anti-HIV compounds.

Other achievements in the ICBG project, in biodiversity conservation initiatives in Vietnam, included the elucidation of plant diversity hotspots at Cuc Phuong National Park (Dietzman et al. 2002), the establishment of the Herbarium of Cuc Phuong National Park (The New York Botanical Garden 2011), the updating of the inventory of the seed plants of Cuc Phuong National Park (Soejarto et al. 2004b), the discovery of plant records new to Cuc Phuong and to Vietnam (Hiep et al. 2002, Soejarto et al. 2002, Bien et al. 2004, Dzu et al. 2004) and of plant taxa new to science (Averyanov and Averyanova 2003), and the establishment of a Threatened Plants Rescue Center at Cuc Phuong National Park. In economic development, included among our accomplishments in Vietnam are the promotion of living standards of members of communities in the Cuc Phuong Commune (Vu et al. 2007), establishment of an ethnomedicinal plant garden in the Cuc Phuong Commune, the documentation of the medicinal plants of the healers of Cuc Phuong (Xuan and Soejarto 2008), the implementation of a "Biodiversity Conservation Support Program," and the strengthening of the research base and capacity of the Cuc Phuong National Park and the Vietnam Academy of Science and Technology. In Laos, our achievements included the establishment of a "Lao Biodiversity Fund" (Lao Biodiversity Fund 2011), the establishment of a medicinal plant database, the establishment of a Medicinal Plant Preserve at Somsavath village of Bolikamsai province (Soejarto et al. 2009), and the completion of a survey of medicinal plants of Laos (Soejarto et al. 2012).

## Perpetuating the Ethnobotany Legacy

The ethnobotany legacy that I inherited from Dick Schultes lives on. In research, the nature of my ethnobotany-related activities has been primarily mission-oriented, in the discovery of non-caloric, high potency sweeteners from plants (Makapugay et al. 1983, Soejarto et al. 1983, Dhammika et al. 1988, Hussain et al. 1990a, 1990b), in the studies of the genus *Stevia* (Asteraceae) (Soejarto 2002), and on a medicinal plant survey of Laos (Libman et al. 2006, 2009; Soejarto et al. 2012).

In teaching, the ethnobotany teachings of Dick Schultes have also continued to live on. At UIC's Graduate Studies in Pharmacognosy (UIC College of Pharmacy 2011), students may elect to graduate in 4 areas of concentration: Natural Products Drug Discovery; Medical Ethnobotany; Biochemical and Molecular Toxicology; and Pharmaceutical Biotechnology. It has been my privilege to serve as Coordinator of the Medical Ethnobotany Track since its establishment in 1997. Since then, I have served as the Major Advisor of six Ph.D. graduates, and as the Co-Advisor of two Ph.D. graduates, in the Medical Ethnobotany Track (Appendix 1).

## ACKNOWLEDGMENTS

I wish to express my humble and deep gratitude to the late Professor Richard Evans Schultes for the guidance he gave me in learning new knowledge during my graduate student years at Harvard University. He gave me the inspiration and the tool to develop an academic career in the search for new medicinal plants to alleviate human sufferings. It has been truly a great privilege and an honor to be a student of Professor Richard Evans Schultes and to inherit and to push on his legacy in the study and teachings of medicinal plants.

Thanks are also expressed to the late Professor Norman R. Farnsworth, a giant of pharmacognosy, whose mentoring has strengthened my confidence to carry out research in the discovery of bioactive compounds from plants as potential candidates for pharmaceutical development.

To my colleagues and members of the Vietnam-Laos ICBG, I wish to express my sincere thanks and appreciation for their support, collaboration, and hard work. Together, we have been able to accomplish the goals we set for this project.

## LITERATURE CITED

Averyanov, L.V., and A.L. Averyanova. 2003. *Updated Checklist of the Orchids of Vietnam.* Vietnam National University Publishing House, Hanoi, pp. 92–96.

Bien, L.K., N.T. Hiep, P.K. Loc, N.M. Cuong, and D.D. Soejarto. 2004. The Diversity of the Flora of Vietnam 14. *Symphorema* Roxb. and *S. involucratum* Roxb., a New Genus and Species Occurence for the Flora of Vietnam. *J Genetics and Applications* 3:36–39.

Bloomberg Businessweek. 2011. As of 12/20/2006, Sarawak MediChem Pharmaceuticals Inc. became a subsidiary of CRAUN Research Sdn. Bhd. Online at http://investing.businessweek.com/research/stocks/private/snapshot.asp?privcapId=8365666. Accessed October 7, 2011.

Boyd, M.R., J.H. Cardellina 2nd, K.R. Gustafson, J.B. McMahon, R.W. Fuller, G.M. Cragg, Y. Kashman, and D.D. Soejarto. 1997. Calanolide and Related Antiretroviral Composition, and Uses Thereof. U.S. Patent 5,591,770. Jan. 7, 1997. Available online at http://www.patents.com/us-5591770.html. Accessed on October 7, 2011.

Case, R. 2005. *Integrative Pharmacognostic Evaluation of Anti-TB Ethnobotanicals from Manus.* Ph.D. Dissertation, Dept. of Medicinal Chemistry and Pharmacognosy, College of Pharmacy, University of Illinois at Chicago.

Corréa S., C.M. 1994. El Depósito de las Plantas: El herbario de la Universidad de Antioquia Cumplió 25 Años de Existencia. *El Espectador,* Wednesday December 7, 1994, p. 4-C. [*El Espectador* is a Colombian Newspaper based in Bogotá. A copy of the article is available from the author upon request.]

Davis, W. 1996. *One River: Explorations and Discoveries in the Amazon Rain Forest.* Simon & Schuster, New York. Chapter 7, p. 209 states: "His precise mandate, however, was to join the hunt for the botanical sources of the arrow and dart poisons known throughout the Amazon as curare. Finding and identifying these plants were of paramount medical importance." Chapter 7, pp. 199–244, and Chapter 10, pp. 296–329 give an account of his visit to the U.S. Embassy in Bogotá and his return to the Amazon. In one of the photographs in Chapter 7, Dick Schultes is shown to score the approximately 650,000 seeds of the rubber plant.

Dhammika, N.P., R.A. Hussain, J.M. Pezzuto, D.D. Soejarto, and A.D. Kinghorn. 1988. An Intensely Sweet Dihydroflavonol Derivative Based on a Natural Product Lead Compound. *J Med Chem* 31:1250–1253.

Dietzman, G., D.V. Khuong, T.Q. Bich, N.M. Cuong, and D.D. Soejarto. 2002. GIS (Geographic Information System)-based Biodiversity Assessment as a Tool to Support Conservation Initiatives at Cuc Phuong National Park. *Advances in Natural Sciences* 3:255–271.

Dzu, N.V., N.T. Hiep, P.K. Loc, and D.D. Soejarto. 2004. Little Known Species and New Records on the Occurrence of Angiosperm Species in Cuc Phuong National Park III. New Records of Aroids (Araceae). *Advances in Natural Sciences* 4:1–14.

Elkington, B.G. 2012 (anticipated). *Anti-Mycobacterial Ethnobotanicals of Laos: Ethnobotany and Pharmacognosy Studies.* Ph.D. Dissertation, Dept. of Medicinal Chemistry and Pharmacognosy, College of Pharmacy, University of Illinois at Chicago.

Farnsworth, N.R. 1984. How Can the Well Be Dry When It Is Filled with Water? *Econ Bot* 38:4–13.

Farnsworth, N.R., W.D. Loub, D.D. Soejarto, G.A. Cordell, M.L. Quinn, and K. Mulholland. 1981. Computer Services for Research on Plants for Fertility Regulation. *Korean J Pharmacol* 12:98–110.

Flavin, M.T., J.D. Rizzo, A. Khilevich, A. Kucherenko, A.K. Sheinkman, V. Vilaychack, L.L. Wei-Chen, E. Mata-Greenwood, T. Pengsuparp, J.M. Pezzuto, S.H. Hugh, T.M. Flavin, M. Cibulski, W. Boulanger, R.L. Shone, and Z.-Q. Xu. 1996. Synthesis, Chromatographic Resolution, and Anti-Human Immunodeficiency Virus Activity of (±)-Calanolide A and Its Enantiomers. *J Med Chem* 39:1303–1313. Available online at http://pubs.acs.org/doi/abs/10.1021/jm950797i. Accessed on October 7, 2011.

Fuller, R.W., H.R. Bokesch, K.R. Gustafson, T.C. McKee, J.H. Cardellina 2nd, J.B. McMahon, G.M. Cragg, D.D. Soejarto, and M.R. Boyd. 1994. HIV-inhibitory Coumarins from Latex of the Tropical Rainforest Tree *Calophyllum teysmannii* var. *inophylloide. Bioorg Med Chem Lett* 4:1961–1964.

Graham, J. 2001. *Cashinahua Medical Ethnobotany.* Ph.D. Dissertation, Dept. of Medicinal Chemistry and Pharmacognosy, College of Pharmacy, University of Illinois at Chicago.

Hamill, F.A. 2001. *Studies of the Medical Ethnobotany of the Buganda Kingdom.* Ph.D. Dissertation, Dept. of Medicinal Chemistry and Pharmacognosy, College of Pharmacy, University of Illinois at Chicago.

Harvard University Herbaria. 2011. Online at http://www.huh.harvard.edu/libraries/Nash/schultes.htm. Accessed on October 7, 2011.

Hiep, N.T., D.D. Soejarto, and P.K. Loc. 2002. The Diversity of the Flora of Vietnam 10. *Nyctocalos* Teijsm. & Binn. (Bignoniaceae), New Genus for the Flora of Vietnam. *Journal of Genetics and Applications* 1: 40–43. [Article in the Vietnamese language.]

Highbeam Business. 2011. Phase II clinical trial was scheduled. Online at http://business.highbeam.com/436989/article-1G1-135758487/calanolide-sarawak-medichem-provides-development-update. Accessed on October 7, 2011.

Hot Stocked. 2011. Advanced Life Sciences (ADLS): Description of Business. Online at http://www.hotstocked.com/companies/a/advanced-life-science-ADLS-description-51106.html. Accessed on October 7, 2011.

Hussain, R.A., L.M. Poveda, J.M. Pezzuto, D.D. Soejarto, and A.D. Kinghorn. 1990a. Phenylpropanoid Constituents of Seven Sweet Tasting Plants. *Econ Bot* 44:174–182.

Hussain, R.A., Y.-M. Lin, L.J. Poveda, E. Bordas, B.S. Chung, J.M. Pezzuto, D.D. Soejarto, and A.D. Kinghorn. 1990b. Plant-derived Sweetening Agents: Saccharide and Polyol Constituents of Some Sweet-Tasting Plants. *J Ethnopharmacol* 28:103–115.

Index Herbariorum. 2011. Online at http://sweetgum.nybg.org/ih/herbarium.php?irn=126256 and http://sciweb.nybg.org/science2/IndexHerbariorum.asp. Accessed on October 7, 2011.

Johnson, H. 2007. *Variability in Bioactive Molecules: Prostratin in* Homalanthus nutans *and BMAA in* Nostoc *and* Cycas micronesica. Ph.D. Dissertation, Dept. of Medicinal Chemistry and Pharmacognosy, College of Pharmacy, University of Illinois at Chicago.

Kahn, E.J., Jr. 1992. Profiles: Jungle Botanist. *The New Yorker*, June 1, pp. 35–58. Available online at http://archives.newyorker.com/?i=1992-06-01#folio=035. Acessed on September 7, 2011 through online purchase. Accessed on October 7, 2011.

Kashman, Y., K.R. Gustafson, R.W. Fuller, J.H. Cardellina II, J.B. McMahon, M.J. Currens, R.W. Buckheit, Jr., S.H. Hughes, G.M. Cragg, and M.R. Boyd. 1992. The Calanolides, a Novel HIV-inhibitory Class of Coumarin Derivatives from the Tropical Rainforest Tree, *Calophyllum lanigerum. J Med Chem* 35:2735–2743.

Koch, A. 2005. *Medical Ethnobotany of the Maasai: In vitro Evaluation of Traditional Anti-malarials.* Ph.D. Dissertation, Dept. of Medicinal Chemistry and Pharmacognosy, College of Pharmacy, University of Illinois at Chicago.

Kreig, M.B. 1964. *Green Medicine: The Search for Plants that Heal.* Rand McNally & Co., Chicago.

Lao Biodiversity Fund. 2011. News about the Lao Biodiversity Fund appeared in the English language version of the Lao Newspaper *Vientiane Times*, January 16, 2007, p. 2. An electronic copy of this news is available from the author upon request.

Libman, A.S. 2006. *Medical Ethnobotany of Laos.* Ph.D. Dissertation, Dept. of Medicinal Chemistry and Pharmacognosy, College of Pharmacy, University of Illinois at Chicago.

Libman, A., S. Bouamanivong, B. Southavong, K. Sydara, and D.D. Soejarto. 2006. Medicinal Plants: An Important Asset to Health Care in a Region of Central Laos. *J Ethnopharmacol* 106:303–311.

Libman, A., B. Southavong, K. Sydara, S. Bouamanivong, C. Gyllenhaal, M.C. Riley, and D.D. Soejarto. 2009. The Influence of Cultural Tradition and Geographic Location on the Level of Medicinal Plant Knowledge Held by Various Cultural Groups in Laos. Pp. 277–291 in C.J. Compton, J.F. Hartman, and V. Sysamouth (eds.), *Contemporary Lao Studies: Research on Development, Culture, Language, and Traditional Medicine, Selected Papers from the First International Conference on Lao Studies, De Kalb, Illinois, May 20–22, 2005.* Center for Laos Studies, San Francisco, and Center for Southeast Asian Studies, Northern Illinois University, De Kalb, Illinois.

Lin, Y.-M., H.M. Anderson, T.R. Jenta, M.J. Williams, M.T. Flavin, and Z.-Q. Xu. 1999. An Efficient and Scalable Method for the Isolation of Costatolide, a Naturally Occurring Anti-HIV Agent, from the Latex of *Calophyllum teysmannii* var. *inophylloide. Pharm Biol* 37:71–76.

Makapugay, H.C., D.D. Soejarto, A.D. Kinghorn, and E. Bordas. 1983. Piperovatine, the tongue-numbing principle of *Ottonia frutescens. J Ethnopharmacol* 7:235–238.

Michel, J. 2006. *Medical Ethnobotany of the Q'eqchi Maya: Perceptions and Botanical Treatments Related to Women's Health.* Ph.D. Dissertation, Dept. of Medicinal Chemistry and Pharmacognosy, College of Pharmacy, University of Illinois at Chicago.

NAPRALERT. 2011. Natural Products Alert: Pharmacology, Biological Activity, Chemistry, Ethnomedicine. Online at http://www.napralert.org/. Accessed on October 7, 2011.

Nguyen, H.T. 2008. *Palatable Prophylaxis Based on Traditional Vietnamese Health Beliefs: An Appealing Approach to Medicine.* Ph.D. Dissertation, Dept. of Medicinal Chemistry and Pharmacognosy, College of Pharmacy, University of Illinois at Chicago, Chicago.

Prance, T.G. 2001. Richard Evans Schultes (12 January 1915 – 10 April 2001): A Tribute. *Econ Bot* 55:347–362.

Sarawak Biodiversity Centre. 2011. The history and present status of Sarawak Biodiversity Centre (SBC) can be found at http://www.sbc.org.my/; http://www.sbc.org.my/en/contact-us; http://www.sarawak.gov.my/en/about-sarawak/education-and-research/research. Accessed on October 7, 2011.

Sarawak MediChem Pharmaceuticals, Inc. 2011. SMP was founded in 1996 and represents a joint venture between the State Government of Sarawak in Malaysia (through its Crop Research & Applications Unit) and MediChem Research. Online at http://pharmalicensing.com/public/companies/view/8440/sarawak-medichem-pharmaceuticals; http://pharmalicensing.com/public/companies/view/8440/sarawak-medichem-pharmaceuticals#company. Accessed on October 7, 2011.

Schultes, R.E. 1947. The Genus *Hevea* in Colombia. *Botanical Museum Leaflets* 12:1–32.

———. 1949. Studies in the Genus *Hevea.* I. The Differentiation of *Hevea microphylla* and *H. minor. Botanical Museum Leaflets* 13:1–11.

———. 1955. The Importance of Plant Classification in *Hevea. Econ Bot* 3:84–88.

———. 1960. The Amazon Indian and Evolution in *Hevea* and Related Genera. *J Arnold Arboretum* 37:123–148.

———. 1967. The Place of Ethnobotany in the Ethnopharmacologic Search for Psychotomimetic Drugs. Pp. 33–57 in D.H. Efron, B. Holmstedt, and N.S. Kline (eds.), *Ethnopharmacologic Search for Psychoactive Drugs.* U.S. Department of Health, Education and Welfare Publication no. 1645. Government Printing Office, Washington, DC.

———. 1972. From Witch Doctor to Modern Medicine: Searching the American Tropics for Potentially New Medicinal Plants. *Arnoldia* 32:198–219.

———. 1978. A New Infrageneric Classification of Hevea. *Botanical Museum Leaflets* 25:243–257.

———. 1986. Conservation of Plant Lore in the Amazon Basin. *Arnoldia* 46:52–59. Available online at http://arnoldia.arboretum.harvard.edu/pdf/articles/707.pdf. Accessed on October 7, 2011.

———. 1991. The Reason for Ethnobotanical Conservation. Pp. 65–75 in O. Akarele, V. Heywood, and H. Synge (eds.), *The Conservation of Medicinal Plants. Proceedings of an International Consultation, 21–27 March, 1988, Chiang Mai, Thailand.* Cambridge University Press, Cambridge, New York.

———. 1993. Plants in Treating Senile Dementia in the Northwest Amazon. *J Ethnopharmacol* 38:129–135.

———. 1994a. The Importance of Ethnobotany in Environmental Conservation. *Am J Econ Sociol* 53:202-206. Available online at http://onlinelibrary.wiley.com/doi/10.1111/j.1536-7150.1994.tb02586.x/abstract. Accessed on September 17, 2011.

———. 1994b. Amazonian Ethnobotany and the Search for New Drugs. Pp. 106–115 in *CIBA Foundation Symposium 185, Ethnobotany and the Search for New Drugs.* John Wiley & Sons, Chichester.

Schultes, R.E., and T. Swain. 1976. The Plant Kingdom: A Virgin Field for New Biodynamic Constitutents. Pp. 134–171 in N.J. Fina (ed.), *The Recent Chemistry of Natural Products, Including Tobacco. Proceedings of the Second Philip Morris Science Symposium, Richmond, Virginia, October 30, 1975.* New York, Philip Morris.

Schultes, R.E., and R.F. Raffauf. 1990. *The Healing Forest: Medicinal and Toxic Plants of the Northwest Amazonia.* Dioscorides Press, Portland, Oregon.

Schultes Images. 2011. Online images of Richard Evans Schultes are at http://web.williams.edu/go/native/schultesobit.htm; http://www.mushroomjohn.org/worldconferenceonshrooms1977b.htm; http://www.huh.harvard.edu/libraries/Nash/schultes.htm; http://www.achievement.org/autodoc/page/sch-3bio-1; http://www.nap.edu/html/biomems/rschultes.pdf; http://www.flickr.com/photos/thinkingarthur/2850951250/; http://erocx1.blogspot.com/2010/07/dr-richard-evans-schultes.html; http://www.flickr.com/photos/thinkingarthur/2850124155/sizes/m/in/set-72157607249988875/; http://www.serendipity.li/dmt/schultes3.htm; http://plants4people.wordpress.com/2009/01/02/the-founder-of-modern-ethnobotany-richard-evans-schultes/; http://skydeviler.hawkdog.net/pub/pix/schultes_aug08_1.jpg; http://news.harvard.edu/gazette/2001/04.19/08-schultes.html; http://www.econbot.org/_about_/index.php?sm=06%7Cawards_schultes; http://elespectador.com/impreso/articuloimpreso136611-manana-llega-el-gringo. Accessed on October 7, 2011.

Soejarto, D.D. 1968. Studies of South American *Saurauia* (Actinidiaceae). Ph.D. Dissertation, Harvard University, Cambridge, Mass.

———. 1980. Revision of South American *Saurauia* (Actinidiaceae). Fieldiana: Botany, New Series No. 2, Publication 1306, 141 pp. Available online at http://www.archive.org/stream/revisionofsoutha02soej/revisionofsoutha02soej_djvu.txt. Accessed on October 7, 2011.

———. 1993. Logistics and Politics in Plant Drug Discovery: The Other End of the Spectrum. Pp. 98–111 in A.D. Kinghorn and M. Balandrin (ed.), *Human Medicinal Agents from Plants.* American Chemical Society, Washington, DC. Available online at http://ukcatalogue.oup.com/product/9780841227057.do. Accessed on October 7, 2011.

———. 2002. Ethnobotany of *Stevia* and *Stevia rebaudiana.* Pp. 40–67 in A.D. Kinghorn (ed.), *Stevia. The Genus Stevia.* Medicinal and Aromatic Plants, Industrial Profiles, Harwood Academic Press, The Netherlands.

Soejarto, D.D., A.S. Bingel, M. Slaytor, and N.R. Farnsworth. 1978. Fertility regulating agents from plants. *B World Health Organ* 56:343–352.

Soejarto, D.D., C.M. Compadre, P.J. Medon, S.K. Kamath, and A.D. Kinghorn. 1983. Potential Sweetening Agents of Plant Origin. II. Field Search for Sweet-tasting *Stevia* Species. *Econ Bot* 37:71–78.

Soejarto, D.D., C. Gyllenhaal, P.S. Ashton, and S.H. Sohmer. 1996. Plant Explorations in Asia Under the Sponsorship of the National Cancer Institute, 1986–1991: An Overview. Pp. 284–310 in M.J. Balick, E. Elisabetsky, and S.A. Laird (eds.), *Medicinal Plant Resources of the Tropical Forest.* Columbia University Press, New York.

Soejarto, D.D., G.M. Cragg, T.C. MacKee, J.H. Cardellina 2nd, M.R. Kadushin, O. Ismawi, H.S. Lee, and M.R. Boyd. 1997. Drug discovery from the tropical rain forests and the conservation of resources: The case of *Calophyllum* (Clusiaceae). Pp. 177–200 in M. Rios and H. Borgtoft (eds.), *Uso y Manejo de Recursos Vegetales: Memorías del Segundo Simposio Ecuatoriano de Etnobotánica y Botánica Económica Oct 1994.* Quito, Ecuador, Abya-Yala Publ. Reprint available from the author upon request.

Soejarto, D.D., C. Gyllenhaal, J.C. Regalado, J.M. Pezzuto, H.S. Fong, G.T. Tan, T.H. Nguyen, T.X. Le, Q.B. Do, V.H. Nguyen, Q.B. Truong, T.H. Nguyen, K.L. Phan, M.V. Bui, B.H. Southavong, K. Sydara, S. Bouamanivong, M.J. O'Neill, J. Lewis, X.M. Xie, and G. Dietzman. 1999. Studies on biodiversity of Vietnam and Laos: A UIC-based ICBG Program. *Pharm Biol* 37(Supplement):1–14.

Soejarto, D.D., N.T. Hiep, P.K. Loc, L.K. Bien, T.D. Dai, J.C. Regalado, and N.M. Cuong. 2002. Little Known Species and New Records on the Occurrence of Angiosperm Species in Cuc Phuong National Park. I. *Heliciopsis lobata, Alangium tonkinense, Illigera parviflora, Keenania tonkinensis, Lysimachia insignis,* and *Kadsura coccinea. Advances in Natural Sciences* 3:57–70.

Soejarto, D.D., C. Gyllenhaal, H.H.S. Fong, L.T. Xuan, N.T. Hiep, N.V. Hung, T.Q. Bich, B. Southavong, K. Sydara, and J.M. Pezzuto. 2004a. The UIC ICBG (University Of Illinois At Chicago International Cooperative Biodiversity Group) Memorandum of Agreement: A Model of Benefit-Sharing Arrangement in Natural Products Drug Discovery and Development. *J Nat Prod* 67(Special issue):294–299.

Soejarto, D.D., N.T. Hiep, P.K. Loc, N.M. Cuong, L.K. Bien, T.D. Dai, J.C. Regalado, M.R. Kadushin, N.T.T. Huong, and T.Q. Bich. 2004b. *Seed Plants of Cuc Phuong National Park: A Documented Checklist.* Published by PCRPS-UIC, Cuc Phuong National Park and Institute of Ecology and Biological Resources, Hanoi.

Soejarto, D.D., H.H.S. Fong, G.T. Tan, H.J. Zhang, C.Y. Ma, S.G. Franzblau, C. Gyllenhaal, M.C. Riley, M.R. Kadushin, J.M. Pezzuto, L.T. Xuan, N.T. Hiep, N.V. Hung, B.M. Vu, P.K. Loc, L.X. Dac, L.T. Binh, N.Q. Chien, N.V. Hai, T.Q. Bich, N.M. Cuong, B. Southavong, K. Sydara, S. Bouamanivong, H.M. Ly, T.V. Thuy, W.C. Rose, and G.R. Dietzman. 2006. "Studies on Biodiversity of Vietnam and Laos" 1998-2005: Examining the Impact. *J Nat Prod* 69:473–481.

Soejarto, D.D., C. Gyllenhaal, J.A. Tarzian Sorensen, H.H.S. Fong, L.T. Xuan, L.T. Binh, N.T. Hiep, N.V. Hung, T.Q. Bich, B.M. Vu, B. Southavong, K. Sydara, J.M. Pezzuto, and M.C. Riley. 2007. Bioprospecting Arrangements: Cooperation Between the North and the South. Pp. 1511–1521 in A. Krattiger, R.T. Mahoney, L. Nelsen, et al. (eds.), *Intellectual Property Management in Health and Agricultural Innovation: A Handbook of Best Practices.* MIHR: Oxford, U.K., and PIPRA: Davis, U.S.A. Available online at www.ipHandbook.org and http://www.iphandbook.org/handbook/ch16/p05/ (Topics #16.5). Accessed on October 7, 2011.

Soejarto, D.D., B. Southavong, K. Sydara, S. Bouamanivong, M.C. Riley, A.S. Libman, M.R. Kadushin, and C. Gyllenhaal. 2009. Studies on Medicinal Plants of Laos: A Collaborative Program between the University of Illinois at Chicago (UIC) and the Traditional Medicine Research Center (TMRC), Vientiane. Pp. 307–323 in C.J. Compton, J.F. Hartman, and V. Sysamouth (eds.), *Contemporary Lao Studies: Research on Development, Culture, Language, and Traditional Medicine. Selected Papers from the First International Conference on Lao Studies, De Kalb, Illinois, May 20–22, 2005.* Center for Lao Studies and the Center for Southeast Asian Studies, Northern Illinois University, DeKalb, Illinois.

Soejarto, D.D., C. Gyllenhaal, M.R. Kadushin, B. Southavong, K. Sydara, S. Bouamanivong, M. Xaiveu, H.J. Zhang, S.G. Franzblau, G.T. Tan, J.M. Pezzuto, M.C. Riley, B.G. Elkington, and D.P. Waller. 2012. An Ethnobotanical Survey of Medicinal Plants of Laos Toward the Discovery of Bioactive Compounds as Potential Candidates for Pharmaceutical Development. *Pharm Biol* 50:42–60.

The Calophyllum Story. 2011. Online at http://www.forestry.sarawak.gov.my/forweb/research/fr/ip/eco/calophys.htm. Accessed on October 7, 2011.

The Duke of Edinburgh Conservation Medal. 2011. Online at http://wwf.panda.org/who_we_are/history/duke_of_edinburgh_conservation_medal/. Accessed on October 7, 2011.

The New York Botanical Garden. 2011. Cuc Phuong National Park [CPNP] Herbarium is listed in the Index Herbariorum. Online at http://sweetgum.nybg.org/ih/herbarium.php?irn=124362 and http://sciweb.nybg.org/science2/IndexHerbariorum.asp). Accessed on October 7, 2011.

The Star Online. 2011. Sarawak Closer to Producing Anti-HIV, Cancer Drugs. Online at http://thestar.com.my/news/story.asp?file=/2009/6/27/nation/20090627195724&sec=nation. Accessed on October 7, 2011.

Tropicos. 2011. Tropicos Nomenclatural Database at http://www.tropicos.org/NameSearch.aspx?name=Micrandra&commonname= and http://www.tropicos.org/NameSearch.aspx?name=Saurauia&commonna=. Accessed on October 7, 2011.

UIC College of Pharmacy. 2011. Graduate Studies in Pharmacognosy at the Department of Medicinal Chemistry and Pharmacognosy, University of Illinois at Chicago. Oline at http://www.uic.edu/pharmacy/depts/Medicinal_Chemistry_and_Pharmacognosy/academics.php. Accessed on October 7, 2011.

Vu, B.M., L.T. Xuan, N.V. Hai, L.X. Dac, D.D. Hai, D.V. Khuong, D.C. Sinh, D.T. Chien, N.T. Nhung, M.C. Riley, C. Gyllenhaal, and D.D. Soejarto. 2007. Drug Discovery, Biodiversity Conservation and Economic Development: A Micro-Loan Program at Cuc Phuong Commune (Vietnam). Pp. 43–62 in J. Schwartz (ed.), *Focus on Biodiversity Research*. Nova Science Publishers, Inc., New York.

Weil, A. 2011. Online at http://www.drweil.com/. Also at https://www.drweilvitaminadvisor.com/drw/ecs/Va2/land_goog_08girl.html?aid=999910&aparam=Google_S_andre_weil&gclid=CJDS95y8kKsCFeoEQAodb0aKyQ. Accessed on October 7, 2011.

Xuan, L.T., and D.D. Soejarto. 2008. *Selected Medicinal Plants of The Muong Community at Cuc Phuong*. Publishing House for Science and Technology, Hanoi. [Book in the Vietnamese language.]

| Year Graduated | Country of field study | Dissertation title |
|---|---|---|
| 2001 | Uganda | Studies of the Medical Ethnobotany of the Buganda Kingdom (Hamill 2001) |
| 2001 | Peru | Cashinahua Medical Ethnobotany and Antimycobacterial Evaluation of Peruvian Plants (Graham 2001) |
| 2005 | Kenya | Medical Ethnobotany of the Maasai: In vitro Evaluation of Traditional Anti-malarials (Koch 2005) |
| 2005 | Papua New Guinea | Integrative Pharmacognostic Evaluation of Anti-TB Ethnobotanicals from Manus (Case 2005) |
| 2006 | Guatemala | Medical Ethnobotany of the Q'eqchi Maya: Perceptions and Botanical Treatments Related to Women's Health (Michel 2006) |
| 2006 | Laos | Medical Ethnobotany of Laos (Libman 2006) |
| 2007 | Samoa | Variability in Bioactive Molecules: Prostratin in *Homalanthus nutans* & BMAA in *Nostoc* & *Cycas micronesica* (Johnson 2007) |
| 2008 | Vietnam | Palatable Prophylaxis Based on Traditional Vietnamese Health Beliefs: An appealing Approach to Medicine (Nguyen 2008) |
| 2012 [expected] | Laos | Anti-Mycobacterial Ethnobotanicals of Laos: Ethnobotany and Pharmacognosy Studies (Elkington 2012) |

Appendix 1. Ph.D. Dissertation of Students Mentored in Pharmacognosy, Area of Concentration in Medical Ethnobotany (University of Illinois at Chicago)

# Whither the Journal Economic Botany? Trends and Transitions, 1947–2010

Robert Voeks,[1,2] Leaa Short,[1] and Aline Gregorio

The scope of the journal *Economic Botany* has changed considerably since its inaugural issue in 1947. In this paper, we explore the nature of these changes, and how they reflect on the field of economic botany/ethnobotany, by means of a review of the 2,187 published research articles, reviews and notes that have appeared since its inception. Contributions transitioned over the decades from largely descriptive reviews of lesser-known crop plants and wild useful plants and fungi to studies of people-plant relations grounded in hypothesis testing. Nearly all articles now include clearly articulated methods and employ quantitative methods. Field research was rarely employed in the early years, but by 2010 was a component in a majority of projects. Early research focused on the United States and neighboring countries, but by 2005 a majority of projects were carried out in less developed countries. The use of laboratory methods peaked in the 1970s, and has held fairly steady at about 25% of articles. One of the clearest trends was the internationalization of the journal, as authorship shifted from primarily North America to include a majority of contributors from Latin America, Asia and Africa. Finally, the content and objectives of submissions changed significantly over time, as researchers increasingly shifted their focus from the real and potential economic value of plants and plant products to ethnobotanical questions—crop plant history and diffusion, ethnobotanical theory, useful species conservation and sustainability, traditional ecological knowledge, the significance of non-timber forest products to local communities, gendered dimensions of people-plant relations, and ethnobotanical knowledge change.

## 1. Introduction

The journal *Economic Botany* has been published continuously as a scholarly quarterly since 1947. The journal was originally published under the auspices of the New York Botanical Garden (NYBG), with Edmond H. Fulling as its first editor-in-chief. Of the eleven editors-in-chief that served over the next 64 years, Richard Evans Schultes holds the distinction of occupying the position for 18 years, by far the longest. In the 1950s, responsibility and content of the journal was transferred to the Society for Economic Botany, where it has remained ever since, although the NYBG maintains the copyright and manages production.

William J. Robbins, Director of NYBG and second editor-in-chief, justified the need for a new journal in its inaugural issue (Robbins 1947). In his article, "Why another botanical magazine?" he argued that a journal of applied botany was needed "to bridge the gap which always exists between [plant] knowledge and its application." (Robbins 1947) There were few if any periodicals at the time devoted to elucidating the vast but under-appreciated utility and economic value of plants to human existence. The new journal would accomplish this, he continued, by publishing information on "plants of potential economic importance, on the less well known methods and facilities by which plants and plant products are used, on problems encountered in utilizing botanical materials, [and] on botanical principles which may be of practical significance . . ." Finally, he noted that the name they had selected for the journal, *Economic Botany*, was "of

1 Department of Geography, California State University—Fullerton, 800 N. State College Blvd., Fullerton, CA 92834. Email: rvoeks@Exchange.FULLERTON.EDU.

2 Editor of *Economic Botany*.

relatively minor importance," and that numerous other names would have been just as valid.

The following year, Raymond Fosburg presented his views on the most promising directions for the journal in "Economic botany—A modern concept of its scope." At its core, he suggested, the field of economic botany is a descriptive endeavor, one that deals with "plants useful or noxious to man, their products, characteristics, uses, geographical origin and distribution" (Fosburg 1948). This descriptive work, which he termed "classical economic botany" was crucial, he argued, for without it, "the more modern and specialized branches [of plant science] cannot exist." He then suggested, however, that the scope of the field should be amplified to include all other dimensions of the plant kingdom that affect or have value to humans, including systematics, forestry, genetics, weed science, and many others. He also discussed the importance of including ethnobotany, which he defined as "the uses of plants by primitive man." Thus, the primary objective of the journal should be to reveal the plant-derived sources of economically useful products and to encourage the search for new products, especially with reference to the knowledge sustained by people in rural communities. Research in these supportive disciplines was necessary, Fosburg argued, in order to revitalize what was "a rather decrepit branch of botany."

More than six decades have elapsed since Fosburg's call to action, and it seems that sufficient time has passed to revisit his charge that economic botany constituted a "decrepit branch of botany." In the following essay, we review the past 64 years of journal articles in *Economic Botany*. We focus on the year-by-year changes in authorship, scope, content, methods and general direction of the discipline.

## METHODS

We reviewed each of the research articles, reviews and notes published since 1947. Because what constituted "research" and "reviews" changed considerably over time, we often used subjective criteria to assign these labels. We recorded the number of articles, reviews, and notes per year, the national origin of primary authors (based on the university/organization address), whether field work was involved and, if so, the country in which it was conducted, the presence or absence of laboratory work, the presence of a clearly-articulated problem and/or hypothesis and methodology, and the use of quantitative methods. We explored the gradual entrance of ethnobotanical topics to complement traditional economic botany work. We also assessed each article in regards to whether or not it included the issue of gendered plant use and knowledge, and whether the article addressed the question of declining knowledge of useful plants.

## RESULTS AND DISCUSSION

In its first 64 years of publication, from 1947–2010, *Economic Botany* published a total of 2,184 research articles, review articles and notes. The first book reviews were published in 1950, and to date there have been a total of 2,681 published. There were also in the early years many hundreds of short, 1–5 paragraph utilization abstracts, mostly excerpts from other botanical and agronomic journals. These were meant to be purely informative and often referred to what for many were novel uses of plants, for example, the estimated economic value of Brazilian tonka beans to the American tobacco industry, the preparation methods of chuño from Andean potatoes, the growth habits of Brazil nuts, and

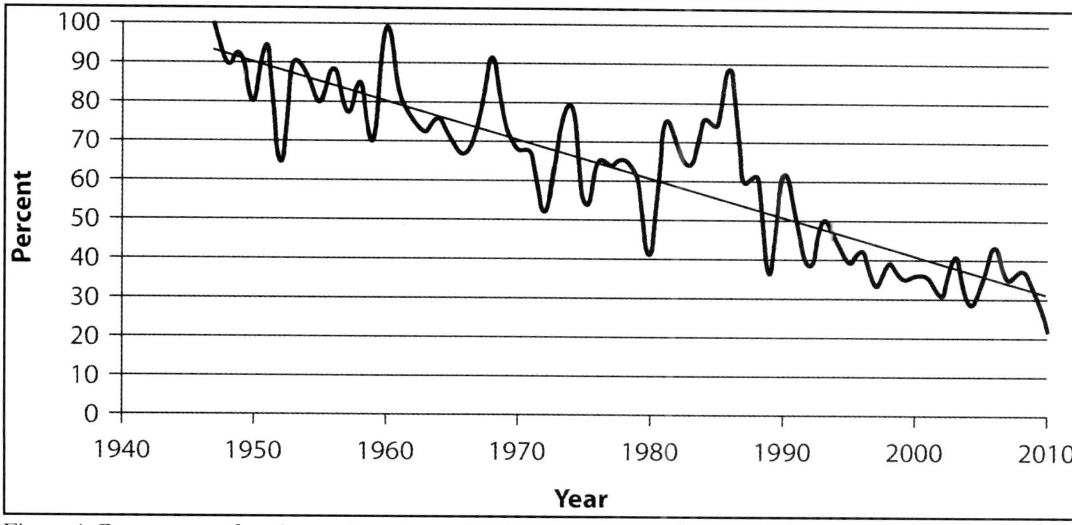


Figure 1. Percentage of authors whose home institution or university was in the United States, 1947–2010 (R = −0.8771; P<0.05).

the like. These were lately replaced by peer-reviewed notes on economically-useful plants.

Among the variables we investigated, a number showed clear and significant trends over the years. The first among these was the startling internationalization of the journal. In the first year of publication, the primary authors were all connected to institutions in the United States (Fig. 1). This changed significantly over the years, however, as more and more authors from Asia, Latin America, Oceania and Africa began to publish in the journal. By the 1980s, North American authors were contributing roughly half of the published manuscripts, and by 2010, this had declined to just over 20%.

The internationalization of the journal can be more accurately understood in terms of the country of origin of the authors. In 1950, for example, contributors hailed from just four countries: the U.S., the U.K., Brazil, and Australia. Over the five year period, 1950–1954, primary authors came from just ten countries, with over 80% being from the U.S (Fig. 2). By contrast, primary authors over the period 2005–2009 came from 33 different countries, and during 2010, only four authors were drawn

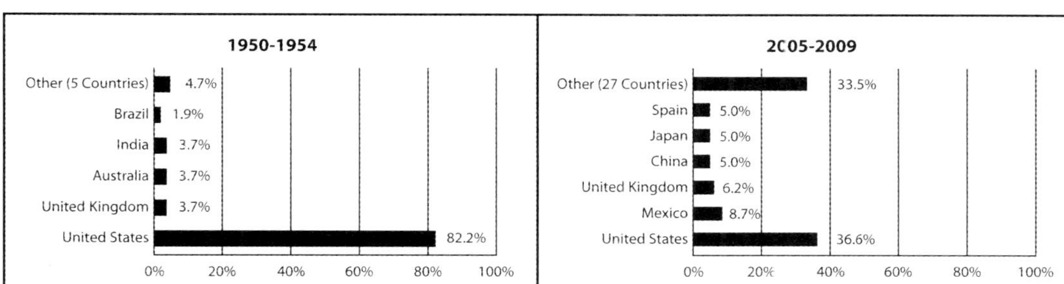


Figure 2. Home institution or university of authors of research articles, reviews and notes, 1950–1954 and 2005–2009.

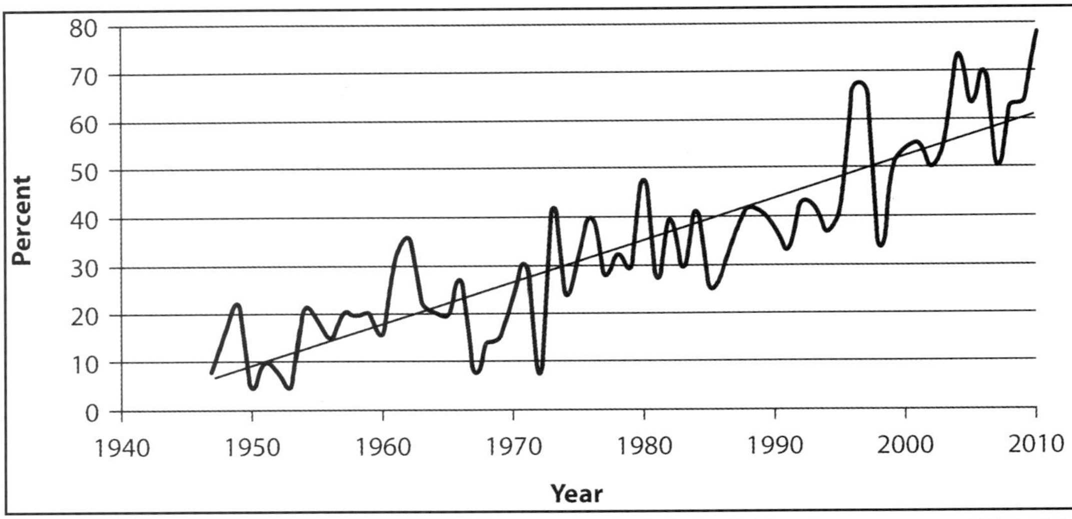


Figure 3. Percentage of articles in Economic Botany that included a fieldwork component, 1947–2010 (R = 0.8736, P<0.05).

from the U.S. In the early decades of publication, research in the field of economic botany and ethnobotany was dominated by American and Western European scientists, but this is clearly not the case today. If publication in *Economic Botany* is a reasonable index of trends in the discipline, scientists from Africa, Asia, and Latin America have emerged as leaders in the study of people-plant relations.

A second feature revealed by this review is an increasing emphasis on fieldwork as opposed to armchair science. During the first few years of publication, few articles included empirical observation and data collection (Fig. 3). They were largely descriptive and informational, really more applied botanical updates than research, and they were often written in a style that appealed as much to the interested lay public as to specialists. For example, in the inaugural year there appeared: *Hybrid Corn—Science in Action* (Morrison 1947); *Manioc—A Tropical Staff of Life* (Schery 1947); *Minor Fiber Industries* (Robinson 1947); and *Veneers and Plywoods—Their Manufacture and Use* (Harrar 1947). There was also a measure of boosterism exhibited by early titles, such as *Ergot—A Blessing and a Scourge* (Youngken 1947), *Growing Better Tobacco* (McMurtrey 1948), and *The Versatile Soybean* (Morse 1947). And this was their objective, for as Robbins (1947) noted, the purpose of the new journal was to remind readers that "without plants. . . man himself could not exist." A fieldwork component in published works increased gradually until the early 1960s, then declined during the late 1960s and early 1970s to levels equivalent to those in the first years of publication. But since the mid-1970s, studies based on field research have increased steadily, and by 2010, nearly 80% of the articles published included fieldwork. With few exceptions, economic botany has become a field science.

The details of this change can be better understood by reviewing the five-year periods 1950–1954 and 2005–2009. In the first five years, barely 10% of studies included a field component, and these were almost wholly North American authors working in the U.S.

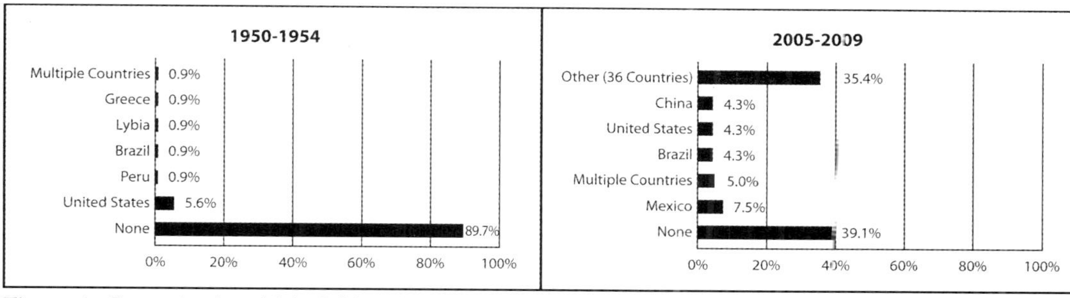


Figure 4. Countries in which fieldwork was carried out in published articles, reviews and notes, 1950–1954, and 2005–2009.

(Fig. 4). By the latter period, however, research in economic botany and ethnobotany was being carried out in the field in over 40 countries. In addition to fieldwork, there has been a trend over the years towards including laboratory work in research articles. This began in the 1960s, and has remained relatively steady at 20% to 30% of articles (Fig. 5).

Perhaps the most strikingly-significant trend the journal exhibited since inception was the transition in research objectives from largely descriptive reviews of crops and wild species to problem-solving projects in the area of people-plant relations. Articles continued to explore the economic utility of crops, drug plants, NTFPs, and other plants and fungi with direct or indirect significance to humans. But beginning in the 1960s, articles increasingly employed hypothesis testing and clearly articulated methodologies to explore questions in economic botany and ethnobotany (Fig. 6). By the mid-1960s roughly 25% of articles described their methods, and this steadily increased over the decades. By 2010, most articles published were constructed around a clearly-stated problem, often presented as a

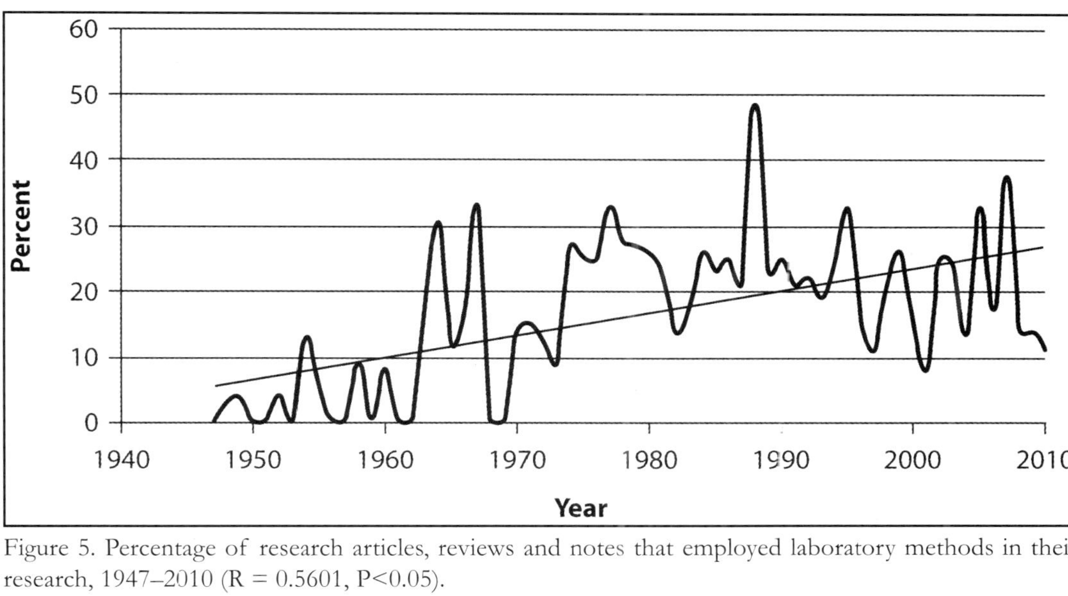


Figure 5. Percentage of research articles, reviews and notes that employed laboratory methods in their research, 1947–2010 (R = 0.5601, P<0.05).

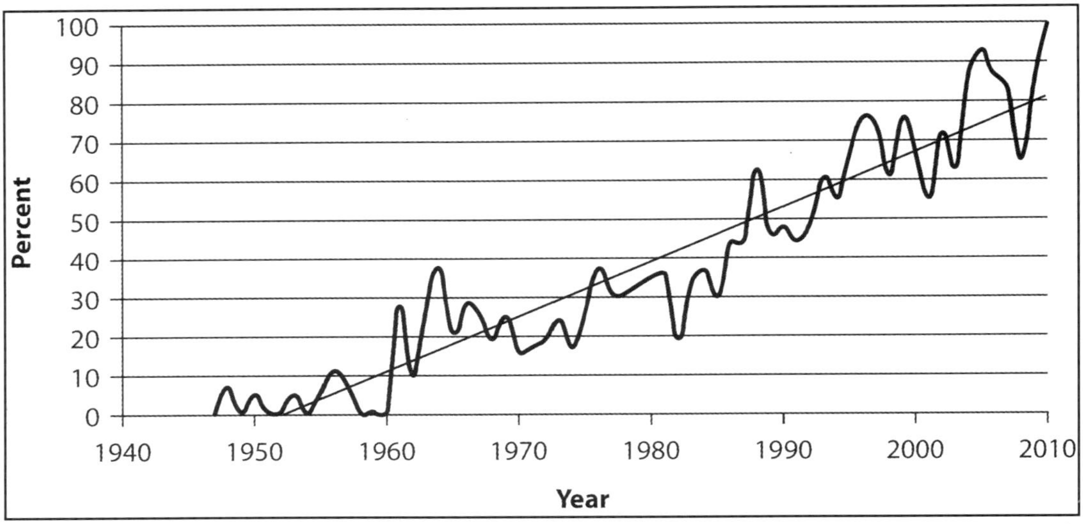


Figure 6. Percentage of articles, reviews and notes that included a methods section, 1947–2010 (R = 0.9400, P<0.05).

hypothesis, and all articles contained a set of objective methods. This trend was equally evident in terms of the use of quantitative methods in experimental design and data analysis (Fig. 7), a trend that began inauspiciously with Gutierrez-Noriega and Wolfgang-Von Hagen's (1951) *Coca—The Mainstay of an Arduous Native Life in the Andes*. The article expounded on the effects of coca leaf addiction, demonstrated that the mean IQ of Andean coca users was 67, and suggested that "80% of coca addicts have sub-normal IQs." As the years

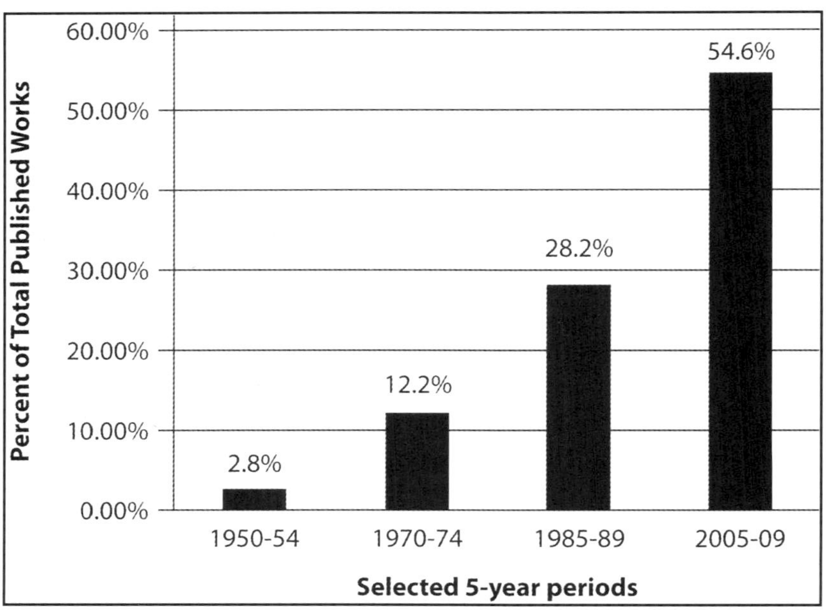


Figure 7. Percentage of published articles, reviews and notes that employed quantitative methods, 1947–2010.

passed, methods increased in rigor and scientists became more mindful of the world of their field informants and field collaborators.

Research continues to extol the value of plants to humankind. Indeed, this remains at the very heart of our enterprise. In the early years, plant values were nearly always framed in terms of distant stakeholders, that is, trade items and the possible development of new products for North Americans and Europeans. For example, when Williams (1961) discussed the need to investigate the ethnobotany of Amazonian trees, he justified this action as "essential to create a domestic and foreign market for these varied products, and to attract capital for the development of forest industries." But the conceptual space encompassing the term "value" has evolved over the decades, and the journal has increasingly served as an outlet for investigations of the myriad means by which local people derive value from plants—social and spiritual as well as economic. What were the ritual values of bombacoids to the ancient Maya (Zidar and Elisens 2009)? Are secondary tropical forests more valuable to healers than primary forests (Voeks 1996)? Do weeds have value to Mexican milpa farmers (Vieyra-Odilon and Vibrans 2001)? What are the subsistence benefits of Brazil's babassu palm (May et al. 1985)? And does collection of agarwood in Laos contribute significantly to the livelihood of local people (Jensen and Meilby 2010)? Value now is defined most often in terms of the people who are immediate contact with nature, those who have the most to lose if these botanical resources are mismanaged or overexploited, and those from whom we have the most to learn about the intimate and nuanced relationship between people and plants.

In spite of Fosburg's call for ethnobotanical research in the journal, this topic did not occur with frequency for several years. From 1947 to 1960, for example, the term "ethnobotany" appeared only 8 times in journal articles, and only one of these, Schultes' *Tapping Our Heritage of Ethnobotanical Lore* (1960) was widely cited. Between 1961 and 1980, the term ethnobotany appeared in 91 articles, and was contained in the title of 21 research articles. Between 1981 and 2000, the term appeared in 481 articles, and was in the title of 96 articles. And between 2001 and 2010, ethnobotany appeared in over 500 articles and notes, or nearly every work that appeared in the journal. Although some of these instances simply recorded bibliographic references with the term in the title, the trend of either doing research that can be considered partly or wholly ethnobotanical in nature, or at least relied in part on ethnobotanical research by others, represents a clear change in the scope of the journal. Indeed, for many researchers and readers, the journal has become principally an outlet for people-plant research.

As *Economic Botany* focused increasingly on ethnobotanical inquiry, several research themes emerged. Among these was the question of declining knowledge of useful species among traditional communities. Researchers have been keenly aware of the fragility of crop and wild-plant knowledge for many years, and globalization of even the most isolated landscapes and people have dramatically accelerated the pace of culture change. Many of these actions have direct impacts on the utility and value of plants to people. This issue was first broached in the journal in 1952 by Muzic (1952), who discussed the impact of the ongoing breakdown in tribal order in Liberia on traditional agricultural practices. Later Clark (1956) reported that the replacement of "native strains" of maize would "de-

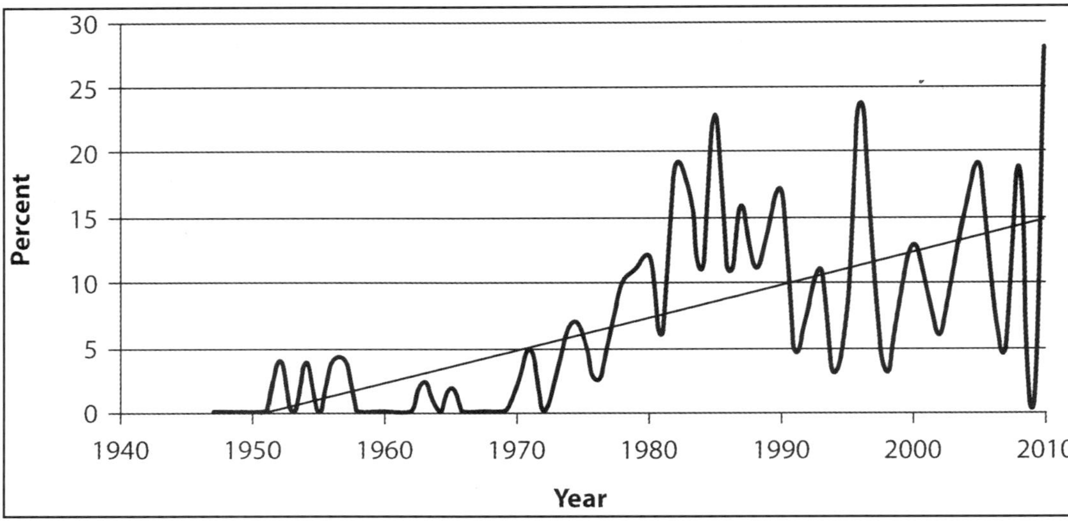


Figure 8. Percentage of articles, reviews and notes that explicitly discussed the process or apparent outcome of declining knowledge of plants, 1947–2010 (R = 0.6631, P<0.05).

prive the rest of the world of genes of great potential importance." And Stopp (1963) was the first to connect language attrition due to acculturation, in this case among indigenous New Guinea speakers, with declining knowledge of medicinal plant species. But cultural erosion and associated loss of plant knowledge was only reported with any regularity beginning in the 1970s (Fig. 8). Weiner (1970) reported, for example, that western influences had so affected the Fiji population that "with the passing of one more generation little of this traditional knowledge will survive." And regarding acculturation in Amazonia, Prance (1973) noted that "it is extremely urgent to gather as much ethnobotanical information as possible before it is permanently lost." Weiss (1979) was one of the first researchers to connect this process to loss of interest by the younger generation, "Sadly many of them [knowledgeable elders] are now dead, often without passing on their knowledge to the young, mainly because the latter are not interested." By the 1980s, about one in ten articles discussed loss of plant knowledge, and from the 1980s to the present, this feature appeared in about 15% of articles. But these early references to landrace and ethnobotanical erosion were in the form of observational narrative, and the first attempt in the journal to quantify the impact of acculturation on useful species knowledge was in Phillips and Gentry's (1993) influential paper on the use of quantitative methods in ethnobotanical research. This was followed by efforts to associate degree of plant knowledge decline or retention with socio-economic variables—age, wealth, geography, formal education, and others—by Benz et al. (2000) in Mexico and by Voeks and Leony (2004) in Brazil). By 2010, over 25% of articles at least mentioned the problem of loss of plant knowledge and skill.

Questions regarding theory in ethnobotany only began to appear in recent years. For example, the issue of possible gendered divisions of plant knowledge, and the implications of divergent knowledge domains, only began to appear in the 1970s, and then only infrequently. The first references were subjective observations of differences between the sexes. The first was by Turner and Bell (1973), who reported that, among British Columbia's Kwakiutl, "As a general rule, the women knew more about plant medicines than the men." Later Latorre and Latorre (1977) compared indigenous use of medicinal plants by Mexico's Kickapoo's with that of local Mestizo women. And Adu-Tutu et al. (1979), in their study of chewing sticks in Ghana, reported that "No difference in amount of collecting was found between males and females, but there are striking differences between the sexes in species selection." The provision of quantitative methods to explore gendered issues in ethnobotany is even more recent. Caniago and Stephen (1996) discovered that older women in Kalimantan, Indonesia possessed significantly more medicinal plant knowledge than men or younger women. And Lozada et al. (2006) explored the possible role of gender in the transmission of knowledge of food and medicinal species in the foothills of the Argentine Andes. Most recently, Reyes-Garcia et al. (2010) reported significant differences in home garden size, aspect, and useful species being cultivated by male and female tenders in Spain.

The relationship between investigator and informant has evolved significantly over the years, particularly after the Convention on Biological Diversity entered into force in 1993. In the early years of the journal, authors treated crop and wild-plant knowledge as open-source information, and informants were often portrayed as rather inconsequential conduits of this information. The idea that plant knowledge and skill represented a form of intellectual property appeared not to have occurred to researchers until the end of the 20th century. In her discussion of the problems of documenting medicinal plants among the hill people of Tibet and Burma, Perry (1961), for example, asserted that local people were "very loath to part with their information. It was as if this knowledge were private property and others had no right to intrude." Indeed, forest dwellers were still being described as "primitive people" in the journal as late as 1990 (Chakrabarty and Rao 1990). In recent years, excepting laboratory or historical work, authors are strongly encouraged during the editorial process to discuss the ethical dimensions of their research in their methods section. These include issues such as prior informed consent, honesty with informants, anonymity, and equitable economic compensation, which are reviewed in The Society for Economic Botany's Guidelines for Professional Ethics (SEB 1995).

Finally, every journal bears the stamp to a greater or lesser degree of its editors. In the case of Richard Evans Schultes, who served as editor-in-chief from January 1962 to December 1979, this influence was considerable. After years of mostly armchair science, *Economic Botany* transitioned from what was a fairly stodgy quarterly to assume in part the image and interests of this great Amazonian botanical explorer. Schultes became the ethnobotanist's archetype, living and working under hostile physical conditions with "primitive" forest folk, plumbing the depth and breadth of their intimate understanding of the plant (and fungal) kingdom, and bringing back to the "civilized" western world newly-discovered botanical treasures for the greater good of society. To a considerable extent, for better or for worse, this image defined the field of economic botany and ethnobotany, and by default the scope of the journal. Consequently, although Schultes did not publish in *Economic Botany* during his long editorial tenure, his preferences as well as his publications established much of the early conceptual breadth of the journal. His pioneering work with hallucinogenic and narcotic plants likewise encouraged numerous publications on this topic during his editorial years—ayahuasca, peyote, salvia, psilocybin, coca, and others—and for some researchers and new-age enthusiasts, the field of ethnobotany was about the quest for hallucinogenic plants. A more important dimension of this legacy is the belief that ethnobotany concerns the discovery of new and useful plants in the tropics, and that these storehouses of genetic and cognitive resources deserve protection. His early articles *Tapping Our Heritage of Ethnobotanical Lore* (Schultes 1960), and later *The Amazonia as a Source of New Economic Plants* (Schultes 1979) stimulated a generation of young ethnobotanists, not to mention a host of novels and Hollywood productions. His later publications in *Economic Botany*, *A Botanical Enigma in the Amazon* (Schultes 1987) and *The Strange Activity of* Malouetia tamaquarina *(Apocynaceae), a Toxic Amazonian Plant* (Schultes 1987), whether intended or otherwise, added a measure of mystery to the emerging field of ethnobotany. Furthermore, the "green heaven" of the lowland tropics were of "incalculable value as an untapped emporium of germplasm for new economic plants" (Schultes 1979), and they above all other biomes merited preservation for the sake of their myriad undiscovered useful plants. Consequently, much of what appeared in *Economic Botany* during the subsequent decades focused on these two themes—ethnobotanical exploration in the tropics and protection of its useful plant resources. For Schultes, the stakes of this ethnobotanical enterprise were enormous, "For the good of our descendants, for the progress of civilization and perhaps even for the survival of the human race, it behooves us—nay, we have the obligation—to protect this nonrenewable gift of Nature"(Schultes 1979).

## REFERENCES


Adu-Tutu, M., Y. Afful, K. Asante-Appiah, D. Lieberman, J. Hall, and M. Elvin-Lewis. 1979. Chewing Stick Usage in Southern Ghana. *Econ Bot* 33:320–328.

Benz, F.F., E.J. Cevallos, M.F. Santana, A.J. Rosales, and M.S. Graf. 2000. Losing Knowledge about Plant Use in the Sierra de Manantlan Biosphere Reserve, Mexico. *Econ Bot* 54:183–191.

Caniago, I., and S. Stephen. 1996. Medicinal Plant Ecology, Knowledge and Conservation in Kalimantan, Indonesia. *Econ Bot* 52:229–250.

Chakrabarty, T., and M. Rao. 1990. A Note on *Glochidion calocarpum* (Euphorbiaceae). *Econ Bot* 44:412–413.

Clark, J. 1956. Collection, Preservation and Utilization of Indigenous Strains of Maize. *Econ Bot* 10:194–200.

Fosburg, F.R. 1948. Economic Botany—A Modern Concept of Its Scope. *Econ Bot* 2:3–14.

Gutierrez-Noriega, C., and V. Wolfgang-Von Hagen. 1951. Coca—The Mainstay of an Arduous Native Life in the Andes. *Econ Bot* 5:145–152.

Harrar, E. 1947. Veneers and Plywoods—Their Manufacture and Use. *Econ Bot* 1:290–305.

Jensen, A., and H. Meilby. 2010. Returns from Harvesting a Commercial Non-Timber Forest Product and Particular Characteristics of Harvesters and Their Strategies: *Aquilaria crassna* and Agarwood in Lao PDR. *Econ Bot* 64:34–45.

Latorre, D., and F. Latorre. 1977. Plants Used by the Mexican Kickapoo Indians. *Econ Bot* 31:340–357.

Lozada, M., A. Ladio, and M. Weigandt. 2006. Cultural Transmission of Ethnobotanical Knowledge in a Rural Community of Northwestern Patagonia, Argentina. *Econ Bot* 60:374–385.

May, P., A. Anderson, M. Balick, and J. Frazão. 1985. Subsistence Benefits from the Babassu Palm (*Orbignya martiana*). *Econ Bot* 39:113–129.

McMurtrey, J. 1948. Growing Better Tobacco. *Econ Bot* 2:326–328.

Morrision, G. 1947. Hybrid Corn—Science in Action. *Econ Bot* 1:5–19.

Morse, J. 1947. The Versatile Soybean. *Econ Bot* 1:137-147.

Muzic, T.J. 1952. Citrus in West Africa, With Special Reference to Liberia. *Econ Bot* 6:246–251.

Perry, L. 1961. Problems in the Compilation of a Native Medicinal Flora in Southeastern Asia. *Econ Bot* 15:241–244.

Phillips, O., and A. Gentry. 1993. The Useful Plants of Tambopata, Peru: II. Additional Hypothesis Testing in Quantitative Ethnobotany. *Econ Bot* 47:33–43.

Prance, G.T. 1972. Ethnobotanical Notes from Amazonian Brazil. *Econ Bot* 26:221–237.

Reyes-García, V., S. Vila, L. Aceituno-Mata, L. Calvet-Mir, T. Garnatje, A. Jesch, J. Lastra, M. Parada, M. Rigat, and J. Vallès. 2010. Gendered Homegardens: A Study in Three Mountain Areas of the Iberian Peninsula. *Econ Bot* 64:235–247.

Robbins, W.J. 1947. Why Another Botanical Magazine? *Econ Bot* 1:3–4.

Robinson, B. 1947. Minor Fiber Products. *Econ Bot* 1:47–56.

Schery, R. 1947. Manioc—Staff of Life. *Econ Bot* 1:20–25.

Schultes, R.E. 1960. Tapping Our Heritage of Ethnobotanical Lore. *Econ Bot* 14:257–262.

———. 1979. The Amazonia as a Source of New Economic Plants. *Econ Bot* 33: 259–266.

———. 1987. A Botanical Enigma in the Amazon. *Econ Bot* 41:454–455

———. 1987. The Strange Activity of *Malouetia tamaquarina* (Apocynaceae), a Toxic Amazonian Plant. *Econ Bot* 41:324–325.

Society for Economic Botany (SEB). 1995. Guidelines for Ethical Research. http://www.econbot.org/pdf/SEB_professional_ethics.pdf

Stopp, K. 1963. Indigenous Plants of the Mt. Hagen People (Mbowamb) in New Guinea. *Econ Bot* 17:16–22.

Turner, N., and M. Bell. 1973. The Ethnobotany of the Southern Kwakiutl Indians of British Columbia. *Econ Bot* 27:257–310.

Vieyra-Odilon, L., and H. Vibrans. 2001. Weeds As Crops: The Value of Maize Field Weeds in the Valley of Toluca, Mexico. *Econ Bot* 55:426–443.

Voeks, R.A. 1996. Tropical Forest Healers and Habitat Preference. *Econ Bot* 50(3):354–373.

Voeks, R.A., and A. Leony. 2004. Forgetting the Forest: Assessing Medicinal Plant Erosion in Eastern Brazil. *Econ Bot* 58(supplement):294–306.

Weiner, M. 1970. Notes on Some Medicinal Plants of Fiji. *Econ Bot* 24:279–282.

Weiss, E.A. 1979. Some Indigenous Plants Used Domestically by East African Coastal Fisherman. *Econ Bot* 33:35–51.

Williams, L. 1961. Natural Wealth of Tropical American Forests. *Econ Bot* 15:223–236.

Youngken, H. 1947. Ergot—A Blessing and a Scourge. *Econ Bot* 1:372–380.

Zidar, C., and W. Elisens. 2009. Sacred Giants: Depictions of Bombacoideae on Maya Ceramics in Mexico, Guatemala and Belize. *Econ Bot* 63:119–129.

# Richard Evans Schultes and the Northwest Amazon

James L. Zarucchi[1]

Richard Evans Schultes was incredibly supportive of his students and provided them opportunities and strong encouragement necessary for successful outcomes – I am one of those lucky people. The following is a brief narrative covering some of my recollections, now several decades old, of Professor Schultes as my professor, mentor, and friend. This is not intended to repeat or summarize his numerous scientific and academic achievements; they are found elsewhere. Many of the observations contained herein are those of a special time when he and I were together collecting plants and information about their uses among people of the Northwest Amazon in Colombia.

I first met Richard Evans Schultes in September 1972 when I enrolled in Biology 104 in the fall of my junior year at Harvard College. The course had been recommended to me by several fellow undergraduates as well as by my Biology advisor, Prof. Martin H. Zimmermann, who was also Director of the Harvard Forest. The course title, *Plants and Human Affairs*, was quite appropriate in that it covered the wide spectrum of plant uses from food, fiber, and building materials to medicines, poisons, narcotics, and hallucinogens. Professor Schultes, who was also the Director of the Botanical Museum, delivered the course lectures in the Nash Lecture Room on the fourth floor of the Museum. The lectures contained much information gained from Schultes' personal experiences, especially those dealing with plants and their uses by people of the Northwest Amazon. Whenever Prof. Schultes would speak about the Amazon, its plants and people, a special smile would appear on his face; deep nostalgia for his time in the Amazon showed clearly. Many other anecdotes about Schultes' style of teaching, such as the proper use of a blowgun in a lecture room, abound in other people's recollections of the great man.

Biology 104 was only one of the four upper-level science courses that I was taking in late 1972, and it was by far the most enjoyable with so many new and interesting things to learn. In some cases, Schultes (always in his white lab coat and red tie) added special twists to his lectures for what might otherwise have been boring topics. The other students in the course were varied, from several first year graduate students to Biology majors wanting an upper-level course to meet departmental requirements, and one seemingly odd student from the humanities who enrolled in the course in greater part because the lecture venue had been named in honor of his grandfather.

Before the semester ended in January 1973, I had decided to change my planned career path to the prospect of doing graduate work with Schultes in Botany rather than pursuing my previous plan of going on to medical school. In addition, I arranged with the help of Profs. Schultes and Zimmermann, for independent study during that spring semester, which allowed me to travel to Colombia and join up with Doel Soejarto; he was a former graduate student of Schultes and at that time Professor of Biology at the Universidad de Antioquia in

[1] Anne L. Lehmann Curator of North American Botany, Missouri Botanical Garden, P.O. Box 299, St. Louis, MO 63166-0299 USA. Email: james.zarucchi@mobot.org.

Figure 1. Richard Evans Schultes (RES) directing Witoto field assistants in the collection of bark from a *Virola* (Myristicaceae) tree near Puerto Nariño, Amazonas, Colombia, March 5, 1975. Photo courtesy of the author.

Medellín. I received very positive encouragement from Prof. Schultes, with him penning letters to his friends and colleagues in Bogotá and Medellín announcing my travel there and asking that they assist me with various matters.

With the support and encouragement of Prof. Schultes, I traveled to Colombia in April and stayed for most of the time (when not in the field or Bogotá) with the Soejarto family in Medellín. I assisted Doel Soejarto on two week-long collecting expeditions to Planta Providencia and environs on the Río Anorí in Antioquia. The first trip was in late April and the other one in the latter part of May, which allowed me to celebrate my 21st birthday in the Colombian jungle. The foray in May also included Schultes' nephew, John Bagley, and Tony Swain of the Royal Botanic Gardens, Kew, there in Colombia testing a field kit for determining the possible presence of alkaloids in fresh plant material.

While I was still in Colombia, Prof. Schultes contacted Doel Soejarto and Isidoro Cabrera (field collaborator with Schultes in the Colombian Amazon for two years in the early 1950s) arranging for them to make some special collections of plants found upriver from Leticia on the main Amazon River at the mouth of the Río Loretoyacú. Luckily, I was able to tag

Figure 2. RES gathering leaves of *Isertia laevis* (Rubiaceae) for chemical studies near Villavicencio, Meta, Colombia, March 23, 1975. Photo courtesy of the author.

along on that collecting trip with those two seasoned botanists.

In spite of various things that happened during my time in Colombia in 1973—such as watching out the window as a motor on a DC-4 stopped functioning (we landed safely back in Leticia), running into a rather large snake one night on the way to the outhouse, and returning home to the US with the larva of two botflies in my back—I was hooked. This profound change in my career path was due totally to my having being provided the opportunity followed by strong support from Schultes and his many friends in Colombia.

Although Prof. Schultes had at first encouraged me as an undergraduate to study *Virola* (Myristicaceae), and that was the subject of my senior thesis, he later suggested, after I started as a graduate student in the fall of 1974, that I find some group within the Apocynaceae for a dissertation topic. My first semester as a graduate student was quite busy with taking classes and co-teaching the weekly Biology 104 laboratory exercise with fellow graduate student Bob Bye. Since much of my graduate student support at that time came from a grant obtained by Prof. Schultes with the clear obligation of obtaining plant samples for chemical studies (based on Schultes' own ethnobotanical observations), I was planning for my 1975 field season in Colombia just as I was starting my first semester as a graduate student. This included my obtaining an initial familiarity with the genera of apocynaceous plants that I might encounter in Amazonia.

I was back in Colombia by the end of February 1975, and was soon joined in Bogotá by Prof. Schultes; we (and some others) would be together in Colombia collecting plants and traveling to interesting locales into early April. I

Figure 3. RES with field assistants collecting inflorescences of *Philodendron dyscarpium* (Araceae) for chemical studies in savanna area near Río Kubiyú, Vaupés, Colombia, March 31, 1975. Photo courtesy of the author.

would remain in Colombia through August of that year. Initially, Schultes organized a collecting trip for us to Leticia and upriver from there to revisit the same general area where I had been in 1973 with Doel Soejarto and Isidoro Cabrera. Schultes and I stayed in Puerto Nariño at the mouth of the Río Loretoyacú and collected with help from local assistants in the nearby forests as well as in degraded areas with pastures. It was wonderful to watch the master ethnobotanist in his element—speaking about the plants and their uses with our assistants using their own Witoto language.

Since Schultes had spent numerous occasions in the region beginning in the early 1940s when he was studying native rubber trees, we were among many plants that he knew well, and I found that he was seeking out "old friends." We went to the place where years earlier he had collected the type specimen of *Ormosia amazonica* var. *venenifera* (Fabaceae); he lamented the fact that the type tree had been cut down, even since I had been there just two years earlier. Also, our attempts to relocate any individuals of *Patinoa ichthyotoxica* (then Bombacaceae, now Malvaceae), where the fruit pulp had been employed by locals as a fish poison, went unrewarded. Even though we tried over various days, no other individuals of either the *Ormosia* or *Patinoa* could be located.

During the days and evenings while we collected in the vicinity of Puerto Nariño, there were many opportunities for us to talk. I learned much more about one of his heroes, a person who had been mentioned numerous times during lectures in Biology 104 and in other courses taught by Schultes. Richard Spruce was an Englishman who spent nearly

15 years in the Amazon and Andes collecting plants and making observations about their distributions and uses. So many of the plants that Schultes had gathered in his many years of collecting in the Northwest Amazon (especially in Colombian Amazonas and Vaupés) had been discovered first by Spruce more than 100 years earlier. Additionally, he spoke of the influence that his professor, Oakes Ames, had on him: that relationship convinced Schultes to pursue graduate studies on plants rather than to proceed on to medical school. And I had encountered that same situation. We also talked about methods that botanists, anthropologists, and other scientists had employed in obtaining ethnobotanical data, and that so much of it was flawed due to the manner in which they interacted with and posed questions to local people. In his interactions with all indigenous people, he was always respectful of their customs and beliefs. After nearly a week in which we gathered numerous specimens for analysis, we returned by boat to Leticia followed by a flight back to Bogotá.

We returned to the Pensión Halifax in the northern part of Bogotá, the residence where Prof. Schultes would normally stay while in the city. Within several days we were joined by Schultes' son Neil and his friend Jimmy Henderson, both teenagers on spring break; they would later accompany us to the Vaupés where Prof. Schultes would help me set up my base of operations in Mitú for collecting efforts in that region of eastern Colombia. Other guests at the pensión included Arthur Sorenson, an Anthropologist/Linguist from

Figure 4. Collecting leaves for chemical studies in savanna area near Río Kubiyú, Vaupés, Colombia, March 31, 1975. Pictured are (left to right) Jimmy Henderson, the author, Neil Schultes, and RES. Photograph taken by Arthur P. Sorenson, Jr.

Figure 5. RES about to deliver a spoonful of coca powder to Neil Schultes as Jimmy Henderson looks on; near Mitú, Vaupés, Colombia, March 1975. Photo courtesy of the author.

SUNY-Binghamton who had been conducting research on multi-lingual Tukanoan people in the Vaupés since the early 1960s—he would also join us later in travel to Mitú.

While we were organizing provisions and equipment for travel to the Vaupés, we made various day-trips out from Bogotá on the road to Villavicencio in order to collect bulk samples of plant material for chemical analysis. One of the plants collected on March 23rd was what we then called *Isertia alba* (Rubiaceae), and later classified by the specialist studying the genus as *Isertia laevis* (Triana) B.M. Boom. The teacher in Professor Schultes never seemed to leave him; we (Neil, Jimmy, Arthur, and I) learned that day, and on other days that followed, much more about the uses of various species of Rubiaceae and many additional plant groups that we encountered. Schultes seemed to be a walking encyclopedia concerning knowledge of native plant uses in Amazonia; he also emphasized the need to always collect quality voucher specimens for any plant encountered where enthobotanical knowledge about them is obtained.

After returning to Bogotá, and three days after collecting the *Isertia*, we went back to Villavicencio by road and made the two-hour flight in a DC-3 from Villavicencio to Mitú. Once there, we made arrangements to start our collecting efforts the very next day when we went the short distance downstream on the Río Vaupés to Urania. Based on some of my later readings, Urania is the place that marks the uppermost ascent on the Rio Negro by famed naturalist Alfred Russel Wallace in

Figure 6. RES and Arthur Sledge at grave of Richard Spruce in England. Photo courtesy of the author.

March 1852. Other forays during our initial time in the Vaupés yielded numerous plants that were completely new to me, and I received proper introductions to them by Prof. Schultes. It was a very rapid learning experience with our botanizing in many distinctly different ecological areas.

One of the Vaupés localities of special interest to Schultes, based on his numerous visits to such places in the Northwest Amazon, was the sandstone (quartzitic) savanna areas upriver from Mitú adjacent to the middle course of the Río Kubiyú. In earlier visits to this locality he had learned from local people of certain uses of plants found on the savannas, and we went there to obtain bulk samples for chemical studies. Among the species that we sought out and collected was *Philodendron dyscarpium*

(Araceae), a species named by Schultes in 1963 based on his collection (with Isidoro Cabrera in April 1953) from a nearby sandstone savanna. We preserved inflorescences of the aroid using alcohol in 5-gallon plastic containers for later study. Schultes related that the species was used in the preparation of an oral contraceptive, and was considered by his informants to be the best of several aroid-based treatments. While still in the area, we made additional collections of bark from various species of *Virola* (Myristacaceae) and gathered leaves of a species later known as *Macrgraviastum mixtum* (Marcgraviaceae) that reportedly had medicinal effect in cases of psychic trauma.

During this time in the Vaupés, Neil Schultes and his friend Jimmy Henderson were shown first-hand by Neil's father many of the places that they had heard about or had only seen in photographs. The teenagers were able to interact with the native population, eat unfamiliar foods such as cassava bread with the local fish-based hot sauce, consume unusual yet savory wild fruits, and partake in activities such as the chewing of powdered coca. Prof. Schultes was incredibly proud of being able to share with his son and Neil's friend even a fraction of the kinds of experiences that he had while residing in the Northwest Amazon; Schultes and the teenagers departed Mitú for home in early April, leaving me to continue my collecting effort.

My introduction to the Vaupés by Schultes helped propel me in my study of apocynaceous plants for my dissertation. I stayed in the region through early August of 1975, and then again I revisited the area during the following year (part of that season was collecting jointly with Mike Balick, another Schultes student, when our focus was on species of palms) and then again for a brief time in 1979. During portions of 1975 and 1976, Arthur Sorenson was able to travel with me and interact with the native population, thus furthering his own research interests. On my part, I was able to observe and collect numerous species in many genera of Apocynaceae, such as *Allamanda*, *Ambelania* (with my dissertation, these became known as species of *Molongum*, *Rhigospira*, *Mucoa*, or *Spongiosperma*, with the latter two described by me as genera new to science), *Aspidosperma*, *Couma*, *Forsteronia*, *Galactophora*, *Himatanthus*, *Lacmellea*, *Macoubea*, *Malouetia*, *Mandevilla*, *Mesechites*, *Odontadenia*, *Parahancornia* (a new species described based on my Vaupés studies), *Prestonia*, *Rhabdadenia*, and *Tabernaemontana*.

The focus of this paper concerned Richard Evans Schultes, the Northwest Amazon, and my relationship with my professor and thesis advisor. I completed my dissertation in late 1981 under the guidance of Schultes, enjoyed two post-doctoral positions (in London and then at the Smithsonian), and then moved to St. Louis where I am currently a Curator at the Missouri Botanical Garden working as the Editorial Director for the Flora of North America project. Throughout this time and up until his passing in 2001, Professor Schultes was supportive, fostered independence, and always provided opportunities, sound advice, and encouragement as my career evolved through the years.